GROUPS AND CHARACTERS

GROUPS AND CHARACTERS

LARRY C. GROVE
Department of Mathematics
University of Arizona
Tucson, Arizona

A Wiley-Interscience Publication
JOHN WILEY & SONS, INC.
New York • Chichester • Weinheim • Brisbane • Singapore • Toronto

Copyright © 1997 by John Wiley & Sons, Inc.

Library of Congress Cataloging in Publication Data:

Grove, Larry C.
    Groups and characters / Larry C. Grove.
        p.   cm. — (Pure and applied mathematics)
    "A Wiley-Interscience publication."
    Includes bibliographical references and index.
    ISBN 0-471-16340-6 (cloth : alk. paper)
    1. Group theory.   I. Title.   II. Series: Pure and applied
mathematics (John Wiley & Sons : Unnumbered)
QA174.2.G77   1997
512'.2—dc21   96-29776
                                                  CIP

Printed in the United States of America
10 9 8 7 6 5 4 3 2 1

Contents

v

Preface

The present volume is intended to be a graduate-level text. It covers some aspects of group theory, concentrating mainly, but not exclusively, on finite groups.

The presentation has been strongly, and positively, influenced by a number of earlier texts and monographs. Particular mention should be made of books by Curtis and Reiner [18], Dornhoff [22], Feit [23], Gorenstein [29], Isaacs [39], and Passman [50].

Chapters 1, 2, and 4 could serve as the text for a basic one-semester course on group theory. Chapter 2 consists entirely of examples, so it could in principal be omitted, but not without radically altering the flavor of the undertaking. Chapter 3 is more easily omitted; on the other hand Chapter 3 is easy and fun, and it provides important applications to combinatorics.

Chapters 5 and 6 contain a basic introduction to ordinary character theory — they do not depend heavily on the preceding four chapters. Chapters 7 through 10 can be read independently of each other in any order following Chapter 6. In fact, Chapter 7 only requires Chapter 5. It should be noted, though, that Chapter 9, on Frobenius groups, makes fairly heavy use of some of the group theory from Chapter 4.

Some attention has been paid to computational aspects of the subject. For example, the Schreier-Sims algorithm, Todd-Coxeter coset enumeration, and various algorithms for calculating character tables are discussed, typically in the context of the very powerful (and free!) computational group theory package GAP.

It is assumed throughout that the reader has assimilated most of the material from a standard first-year graduate abstract algebra course in a U.S. university, such as in [31]. This includes elementary group theory, such as Sylow theorems (although a proof is included in Chapter 1), presentations, solvability and nilpotence, etc.; as well as basic facts about rings, modules, and field extensions. It is important for the reader to have a reasonable facility with linear algebra. Nevertheless, some basic linear algebra is included in the text, on the grounds that it may not always be covered in standard undergraduate courses.

I wish to thank the faculty, students, and staff of Lehrstuhl D für Mathematik at the RWTH in Aachen for support and stimulation during two sabbatical leaves. Particular acknowledgment is in order for Professor Joachim Neubüser, the founding father of GAP; it was my privilege to attend many of his wonderfully clear lectures on groups and representations.

My special thanks to Robert Beals and Olga Yiparaki for careful readings of parts of the manuscript.

Larry C. Grove

The University of Arizona
August 1996

GROUPS AND CHARACTERS

Chapter 1

Preliminaries

In this chapter we present a variety of concepts and ideas, in part to establish terminology, usage, and notation. Everything presented will appear in later chapters.

It is assumed that the reader is familiar with the elementary group-theoretical material normally covered in a standard first graduate-level algebra course (in the United States), and also the material from a junior/senior-level linear algebra course. Any gaps can be filled by browsing in one or more of the many texts available for such courses.

1.1 Some Notation and Generalities

If G is a group and $x, y \in G$ we shall write x^y to denote $y^{-1}xy$ and ^yx to denote yxy^{-1}. Note that $x^{yz} = (x^y)^z$ and $^{yz}x = {}^y({}^zx)$ for all $x, y, z \in G$.

If G is a group and $H \subseteq G$ is a subgroup we write $H \leq G$, or $G \geq H$, as usual. If $H \leq G$ then a set T of (right) coset representatives for H in G will be called a (right) *transversal*; we will **always** assume that $T \cap H = 1$, i.e. that the representative of H itself is the identity element. Note that $|T| = [G : H]$, the index of H in G.

If T is a transversal for H in G we define a "transversal function" from G to T, denoted $x \mapsto \bar{x}$, via $\bar{x} = t$, where $t \in T$ and $Hx = Ht$, or equivalently $Hx \cap T = \{\bar{x}\}$. Note that the notation depends completely on the choice of a fixed transversal T; if S is *any* transversal then $\{\bar{s} : s \in S\} = T$.

The first proposition lists three trivial but important facts about the transversal function.

Proposition 1.1.1 *Suppose that T is a transversal for $H \leq G$. Then*

1. *$\bar{\bar{x}} = \bar{x}$,*

2. *$\bar{x}x^{-1} \in H$, and*

1

3. $\overline{xy} = \overline{\overline{x}y}$

for all $x, y \in G$.

Proof Parts 1 and 2 are obvious. For part 3 note that $H\overline{xy} = Hxy = (Hx)y = H\overline{x}y$, so $\overline{\overline{x}y} = \overline{xy}$.   △

Corollary 1.1.2 *If* $t \in T$ *and* $x \in G$ *then* $\overline{\overline{txx^{-1}}} = t$.

Proof $\overline{\overline{txx^{-1}}} = \overline{txx^{-1}} = \overline{t} = t$.   △

Suppose now that $H \leq G$ and $G = \langle X \rangle$, i.e. that X is a set of generators for G. Write $X^{-1} = \{x^{-1} : x \in X\}$ and $X^{\pm} = X \cup X^{-1}$. Suppose that T is a transversal for H in G, and set

$$A = \{tx\overline{tx}^{-1} : t \in T,\ x \in X^{\pm}\} \subseteq H,$$

$$B = \{tx\overline{tx}^{-1} : t \in T,\ x \in X\} \subseteq A.$$

Proposition 1.1.3 *In the setting above* $A \subseteq B \cup B^{-1} \cup \{1\}$.

Proof Take $t \in T$ and $x \in X$. It will be sufficient to show that $tx^{-1}\overline{tx^{-1}}^{-1} \in B^{-1}$. Set $t_1 = \overline{tx^{-1}} \in T$. Then $t_1 x\overline{t_1 x}^{-1} \in B$, and $(t_1 x\overline{t_1 x}^{-1})^{-1} = \overline{t_1 x}x^{-1}t_1^{-1} = \overline{\overline{tx^{-1}}x}x^{-1}\overline{tx^{-1}}^{-1} = tx^{-1}\overline{tx^{-1}}^{-1} \in B^{-1}$.   △

Corollary 1.1.4 $\langle A \rangle = \langle B \rangle$.

Bear in mind that if T is a transversal for H in G then it is always assumed that $T \cap H = 1$.

Theorem 1.1.5 (Schreier) *Suppose that* G *is a group,* $G = \langle X \rangle$, $H \leq G$, *and* T *is a transversal for* H *in* G. *Then* H *is generated by the set*

$$B = \{tx\overline{tx}^{-1} : t \in T,\ x \in X\}.$$

Proof By Corollary 1.1.4 it will suffice to show that $H = \langle A \rangle$, with $A = \{tx\overline{tx}^{-1} : t \in T,\ x \in X^{\pm}\}$. For any $y \in H$ write $y = x_1 x_2 \cdots x_k$, with each $x_i \in X^{\pm}$. Set $t_1 = 1$; then $t_2 = \overline{t_1 x_1}$, $t_3 = \overline{t_2 x_2}$, ..., $t_{k+1} = \overline{t_k x_k} \in T$. Next set

$$a_i = t_i x_i \overline{t_i x_i}^{-1} = t_i x_i t_{i+1}^{-1} \in A,$$

$1 \leq i \leq k$, and observe that

$$\begin{aligned}
a_1 a_2 \cdots a_k &= (t_1 x_1 t_2^{-1})(t_2 x_2 t_3^{-1}) \cdots (t_k x_k t_{k+1}^{-1}) \\
&= t_1 x_1 x_2 \cdots x_k t_{k+1}^{-1} = 1 \cdot y t_{k+1}^{-1} \in H.
\end{aligned}$$

But then $t_{k+1} \in H$, since $y \in H$, and also $t_{k+1} \in T$, so $t_{k+1} = 1$, i.e. $y = a_1 a_2 \cdots a_k \in \langle A \rangle$.   △

Corollary 1.1.6 *If G is a finitely generated group, $H \leq G$, and $[G:H]$ is finite, then H is finitely generated.*

Proof Say $G = \langle X \rangle$, with X finite, and let T be a transversal for H in G. Then $H = \langle B \rangle$, as in the theorem, and $|B| \leq |T||X| = [G:H]|X|$. $\triangle$

By way of contrast if G is the free group on two generators and $H = G'$, the commutator subgroup, then H is known to be free on countably infinitely many generators, so $[G:H]$ must be infinite.

The set $B = \{tx\overline{tx}^{-1} : t \in T,\ x \in X\}$ is called the set of *Schreier generators* for $H \leq G$, relative to the generating set X and transversal T.

For an example take G to be the symmetric group S_n and H to be the alternating subgroup A_n. Let σ be the transposition $(1\,2)$ and τ the n-cycle $(1\,2 \cdots n)$. Then S_n is generated by $X = \{\sigma, \tau\}$, and $T = \{1, \sigma\}$ is a transversal for A_n in S_n. Assume that n is odd, so $\tau \in A_n$. Schreier generators then are $1\sigma\overline{1\sigma}^{-1} = \sigma\sigma^{-1} = 1$, $1\tau\overline{1\tau}^{-1} = \tau$, $\sigma\sigma\overline{\sigma\sigma}^{-1} = 1$, and $\sigma\tau\overline{\sigma\tau}^{-1} = \sigma\tau\sigma$, so $A_n = \langle \tau, \sigma\tau\sigma \rangle$. Note that $\sigma\tau\sigma = (2\,1\,3 \cdots n)$. What are the Schreier generators for A_n if n is even?

1.2 Permutation Actions

If S is a set write $\mathrm{Perm}(S)$ to denote the set of all permutations of S, i.e. all bijections $\sigma : S \to S$. It will usually be convenient to use (right) "exponential" notation for permutation actions: if $s \in S$ and $\sigma \in \mathrm{Perm}(S)$ then s^σ is the image of s under the action of σ.

If $\sigma, \tau \in \mathrm{Perm}(S)$ we define their composition product $\sigma\tau$ as usual: $s^{\sigma\tau} = (s^\sigma)^\tau$, all $s \in S$. Note that with exponential notation σ acts first, followed by τ, in $\sigma\tau$, i.e. multiplication is left-to-right. With composition as the product $\mathrm{Perm}(S)$ is a group.

Very often S will be the set $\{1, 2, \ldots, n\}$, in which case we will as usual call $\mathrm{Perm}(S)$ the *symmetric group* and denote it by S_n or $\mathrm{Sym}(n)$. Furthermore we will use the usual cycle notation, etc., bearing in mind that multiplication proceeds from left to right.

If S is a set and G is any group we say that G *acts on* S if there is a homomorphism $\theta : G \to \mathrm{Perm}(S)$, in which case for $s \in S$ and $x \in G$ we have $s \mapsto s^{\theta x}$. If θ is one-to-one the action is called *faithful*. In most situations the θ will be suppressed, and we will simply write $s \mapsto s^x$. With that convention the defining properties of a group action are $s^1 = s$ and $s^{xy} = (s^x)^y$, all $s \in S$ and all $x, y \in G$.

The cardinality $|S|$ is often called the *degree* of the action.

In some situations the exponential notation for group actions can be confused with, or even conflict with, other established notations. To avoid such problems we will usually write instead $s \mapsto sx$ (or in some cases $s \mapsto {}^x s$, or $s \mapsto xs$), with obvious corresponding changes in the governing properties.

If G acts on S and $s \in S$ recall that the *orbit* of s is $\mathrm{Orb}_G(s) = \{s^x : x \in G\}$, and that the *stabilizer* of s is $\mathrm{Stab}_G(s) = \{x \in G : s^x = s\} \leq G$. Alternate notations that will sometimes be useful are s^G for $\mathrm{Orb}_G(s)$ and G_s for $\mathrm{Stab}_G(s)$.

Exercise

Suppose that G acts on S, $x \in G$, and $s \in S$. Show that $G_{s^x} = G_s^x$, i.e. stabilizers of elements in the same orbit are conjugate.

Recall also that G is said to act *transitively* on S if $\mathrm{Orb}_G(s) = S$ for some (hence all) $s \in S$.

If G_1 acts on S_1 and G_2 acts on S_2 we say that the actions are *equivalent* if there is an isomophism $\varphi : G_1 \to G_2$, together with a bijection $f : S_1 \to S_2$ such that the diagram

$$
\begin{array}{ccc}
S_1 & \xrightarrow{\ x\ } & S_1 \\
f \downarrow & & \downarrow f \\
S_2 & \xrightarrow{\varphi(x)} & S_2
\end{array}
$$

is commutative; i.e. $f(s^x) = f(s)^{\varphi(x)}$ for all $x \in G_1$, $s \in S_1$.

Exercises

1. If $H \leq G$ show that G acts transitively on the set $\{Hy\}$ of all right cosets of H in G, with the action of $x \in G$ being given by $x : Hy \mapsto Hyx$.

2. Show that the kernel of the action in exercise 1 is $\cap\{H^x : x \in G\}$, which is called the *core* of H in G and denoted $\mathrm{core}(H)$.

3. Show that $\mathrm{core}(H)$ is the largest subgroup of H that is normal in G.

4. If G acts transitively on S and $s \in S$ show that the action of G on S is equivalent with the action (above) of G on the set of cosets of the stabilizer $H = \mathrm{Stab}_G(s)$.

The next proposition is a slight variation on the familiar (and very basic) fact that the size of an orbit is the index of the stabilizer.

Proposition 1.2.1 *Suppose that G acts on S, $s \in S$, and T is a transversal for $\mathrm{Stab}_G(s)$ in G. Then the map $\varphi : t \mapsto s^t$ is a bijection from T to $\mathrm{Orb}_G(s)$.*

Proof Check that φ is 1–1 and onto.   △

Let us illustrate the proposition by means of an example. Set $\sigma = (1\,2\,3\,4\,5)$, $\tau = (2\,3\,5\,4)$, and $G = \langle \sigma, \tau \rangle \leq S_5$. Then $1^\sigma = 2$, $1^{\sigma\tau} = 2^\tau = 3$, $1^{\sigma\tau\sigma} = 3^\sigma = 4$, and $1^{\sigma\tau^2} = 3^\tau = 5$, so G is transitive, $\mathrm{Orb}_G(1) = S = \{1, 2, 3, 4, 5\}$. Thus $G_1 = \mathrm{Stab}_G(1)$ has index 5 in G.
"Graphically" we have

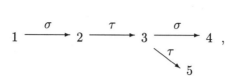

and furthermore we have a transversal $T = \{1, \sigma, \sigma\tau, \sigma\tau\sigma, \sigma\tau^2\}$ for G_1 in G. This is in fact an example of a *Schreier transversal*, characterized by the fact that each element of T is obtained from an earlier element of T via multiplication (on the right) by one of the given generators for G.

Exercise

Carry out the same procedure for the subgroup G of S_7 generated by $\sigma = (2\,4)(5\,6)$ and $\tau = (1\,2\,5\,7)(3\,4)$, and find a transversal for G_1 in G.

We shall pursue the ideas above somewhat further in order to sketch an algorithm for determining the order of a (finite) permutation group.

If G acts on S and $s_1, s_2 \in S$ write G_{s_1,s_2} for $G_{s_1} \cap G_{s_2}$, the *2-point stabilizer*, and similarly for 3-point stabilizers, etc. In general, if $T \subseteq S$ write G_T to denote $\cap\{G_t : t \in T\}$, the subgroup that simultaneously stabilizes all points of T.

A sequence $(b_1, b_2, \ldots, b_k)$ of distinct points from S is called a *base* for G if $G_{b_1, b_2, \ldots, b_k} = 1$.

For example, if $G = S_4$ then $B = (1, 2, 3)$ is a base; if $G = \langle (1\,3), (1\,2\,3\,4) \rangle$, a dihedral subgroup of S_4, then $B = (1, 2)$ is a base.

Proposition 1.2.2 *Suppose that $B = (b_1, \ldots, b_k)$ is a base for G, x, $y \in G$, and $b_i^x = b_i^y$, $1 \le i \le k$. Then $x = y$. Thus elements of G are determined by their action on a base.*

Proof We have $b_i^{xy^{-1}} = b_i$, all i, so $xy^{-1} \in G_{b_1,\ldots,b_k} = 1$ and $x = y$. △

The usefulness of the proposition derives mainly from the fact that surprisingly often even large groups of high degree have bases that are rather short.

Now for some slightly elaborate notation. Suppose G acts on S, and that $B = (b_1, \ldots, b_k)$ is a base for G. Write $G^{(1)} = G$, $G^{(2)} = G_{b_1}^{(1)} = G_{b_1}$, $G^{(3)} = G_{b_2}^{(2)} = G_{b_1,b_2}$, etc., then $\mathcal{O}_1 = b_1^{G^{(1)}}$, $\mathcal{O}_2 = b_2^{G^{(2)}}$, etc. Let T_i be a transversal for $G^{(i+1)}$ in $G^{(i)}$, $i = 1, \ldots, k$. Note that $G^{(k+1)} = G_{b_1,b_2,\ldots,b_k} = 1$. Note also that $|T_i| = |\mathcal{O}_i|$, all i, by Proposition 1.2.1.

Proposition 1.2.3 *With the notation and setting above each $x \in G$ can be written uniquely as $x = t_k \cdots t_2 t_1$, with $t_i \in T_i$, all i.*

Proof Let $t_1 = \bar{x} \in T_1$ (relative to $G^{(2)} \leq G^{(1)}$) and write $x = x_2 t_1$ (uniquely), with $x_2 \in G^{(2)}$. Then let $t_2 = \overline{x_2} \in T_2$ and write $x_2 = x_3 t_2$ (uniquely), with $x_3 \in G^{(3)}$, hence $x = x_3 t_2 t_1$, etc. At stage k we have $x = x_{k+1} t_k \cdots t_1$, and $x_{k+1} \in G^{(k+1)} = 1$. $\triangle$

Corollary 1.2.4 $|G| = \prod_{i=1}^{k} |T_i| = \prod_{i=1}^{k} |\mathcal{O}_i|.$

Two questions arise naturally. How can we find a base B for G? Once we have B, how can we determine the subgroups in the stabilizer chain $G^{(2)} \geq G^{(3)} \geq \cdots \geq G^{(k)}$?

One more definition. If B is a base for the group G acting on a set S, then a *strong generating set* is a set X of generators for G such that $\langle X \cap G^{(i)} \rangle = G^{(i)}$ for $1 \leq i \leq k$.

We return to the example on page 4 to illustrate how to obtain a base B and strong generating set X. The basic idea is to add elements to B one at a time, get a transversal via Proposition 1.2.1, calculate Schreier generators $tx\overline{tx}^{-1}$ for the stabilizer, and continue until a stabilizer is trivial.

Recall that $G = \langle \sigma, \tau \rangle$, with $\sigma = (1\,2\,3\,4\,5)$ and $\tau = (2\,3\,5\,4)$. It was determined that $\mathcal{O}_1 = S = \{1, 2, 3, 4, 5\}$, and we have a transversal $T_1 = \{1, \sigma, \sigma\tau, \sigma\tau\sigma, \sigma\tau^2\}$ for $G_1 = G^{(2)}$ in G. We begin with $B = (1)$ and $X = \{\sigma, \tau\}$.

The table below shows the Schreier generators for G_1.

t	x	$\overline{tx}$	$tx\overline{tx}^{-1}$
1	σ	σ	1
σ	σ	$\sigma\tau$	τ^{-1}
$\sigma\tau$	σ	$\sigma\tau\sigma$	1
$\sigma\tau\sigma$	σ	$\sigma\tau^2$	τ^{-1}
$\sigma\tau^2$	σ	1	τ^2
1	τ	1	τ
σ	τ	$\sigma\tau$	1
$\sigma\tau$	τ	$\sigma\tau^2$	1
$\sigma\tau\sigma$	τ	σ	τ^2
$\sigma\tau^2$	τ	$\sigma\tau\sigma$	τ^2

How are the calculations done? For example, in the second line of the table we have $tx = \sigma^2$, which is not a transversal element. But $\sigma^2 = (1\,3\,5\,2\,4)$, which carries 1 to 3, as does the transversal element $\sigma\tau$; hence $\overline{\sigma^2} = \sigma\tau$. The Schreier generator is thus $\sigma^2 \tau^{-1} \sigma^{-1}$, which is τ^{-1}.

We see then that $G^{(2)} = \langle \tau \rangle$. Note that $X \cap G^{(2)} = \{\tau\}$, which generates $G^{(2)}$, so there is no need to enlarge X.

We proceed by adding another point to B, say $B := (1, 2)$. For $\mathcal{O}_2$ and T_2 we may use

$$2 \xrightarrow{\ \tau\ } 3 \xrightarrow{\ \tau\ } 5 \xrightarrow{\ \tau\ } 4 \,,$$

so $\mathcal{O}_2 = \{2, 3, 5, 4\}$ and $T_2 = \{1, \tau, \tau^2, \tau^3\}$, a transversal for $G^{(3)} = G_{1,2}$ in $G^{(2)}$.

Easy calculations show that the Schreier generators for $G^{(3)}$ are all trivial. Thus $G^{(3)} = 1$, and it is clear that $B = (1, 2)$ is a base for G with strong generating set $X = \{\sigma, \tau\}$. Furthermore we see that $|G| = |T_1||T_2| = 5 \cdot 4 = 20$.

The procedure sketched above is called the *Schreier-Sims* algorithm.

The algorithm as sketched has the potential for producing large numbers of generators, with adverse effects on efficiency. There is a simple observation, however, that allows determination of a reasonable bound on $|X|$. Suppose e.g. that $G \leq S_n$ has base $B = (b_1, b_2, \ldots)$. Suppose that σ_1 and σ_2 are generators that move b_1, and that $b_1^{\sigma_1} = b_1^{\sigma_2}$. Then σ_1 and σ_2 can be replaced by σ_1 and $\sigma_2\sigma_1^{-1}$, and $\sigma_2\sigma_1^{-1}$ stabilizes b_1. When all such replacements are made there can be at most $n - 1$ generators that actually move b_1. Similar replacements assure at most $n - 2$ generators that stabilize b_1 and move b_2, etc. Thus as we move down the stabilizer chain we see that we may keep

$$|X| \leq (n - 1) + (n - 2) + \cdots + 2 + 1 = \frac{n^2 - n}{2}.$$

Exercises

1. Apply the Schreier-Sims algorithm to the subgroup of S_6 generated by $(1\,3)(4\,6)$ and $(1\,2\,5)$. You should find a base of length 3, and the original generating set will have to be enlarged to include generators for stabilizers. Find $|G|$.

2. Apply the algorithm to the subgroup G of S_7 in the exercise on page 5.

As the exercises above show it soon becomes rather tedious to carry out the algorithm by hand. Fortunately a sophisticated version of it has been implemented in the group theoretical software package GAP [57]. The symbolic algebra package Maple [12] also has rudimentary permutation group operations available, including calculations of orders, which presumably also relies on an implementation of the Schreier-Sims algorithm.

Here is a brief sample of a GAP session, during which the Schreier-Sims algorithm is carried out in the background.

```
gap> G := Group((1,2,3,4,5,6,7,8,9,10,11),
    (3,7,11,8)(4,10,5,6));
    Group((1,2,3,4,5,6,7,8,9,10,11),(3,7,11,8)(4,10,5,6))
```

```
gap> Size(G);
  7920
```

The group in the example will reappear in Chapter 2.

GAP is implemented for most commonly used computer platforms, such as Unix, DOS, and Macintosh. It is available at no charge from a number of sources in Europe and the United States. Readers are urged to search for it on the World Wide Web, to download it, and to use it.

Exercise

Use GAP to find the order of the subgroup of S_{12} generated by the two permutations in the example above together with the permutation

$$(1, 12)(2, 11)(3, 6)(4, 8)(5, 9)(7, 10).$$

The Sylow theorems for finite groups can be proved simultaneously by means of appropriate group actions. Recall that if p is a prime and p^k is the exact power of p dividing the order of a group G, then a Sylow p–subgroup of G is any subgroup of order p^k. Write n_p for the number of distinct Sylow p-subgroups in G.

Theorem 1.2.5 (L. Sylow) *Suppose that G is a finite group, $p \in \mathbb{N}$ is a prime, and $|G| = p^k m$, with $p \nmid m$. Then*

S1: Sylow p–subgroups exist,

S2: any 2 Sylow p–subgroups are conjugate in G, and

S3: $n_p \equiv 1$ (mod p).

Proof We may assume $k \geq 1$. Denote by $\mathcal{P}$ the set of all p–subgroups of G that are maximal relative to inclusion. Then G acts on $\mathcal{P}$ by conjugation, as does any subgroup of G. Note that if $P \neq Q$ in $\mathcal{P}$ then $Q^P \neq Q$ (i.e. $\mathrm{Orb}_P(Q) \neq \{Q\}$), or else $P < PQ$, a p–subgroup since $PQ/Q \cong P/P \cap Q$, contradicting the maximality of P. Fix $P \in \mathcal{P}$ and let $\mathcal{O}$ denote $\mathrm{Orb}_G(P)$. Then P is the only fixed point in the action of P on $\mathcal{O}$, so $|\mathcal{O}| \equiv 1$ (mod p). If $Q \in \mathcal{P} \setminus \mathcal{O}$ then there are no fixed points in the action of Q on $\mathcal{O}$, so $|\mathcal{O}| \equiv 0$ (mod p), a contradiction, and hence $\mathcal{O} = \mathcal{P}$. Thus S2 and S3 are true if S1 is true.

As for S1, if $|P| < p^k$ then $p \mid [N_G(P) : P]$ (since $[G : N_G(P)] = |\mathcal{P}| \equiv 1$ (mod p)), and so $N_G(P)/P$ has a subgroup L/P of order p by Cauchy's Theorem. But then $|L| = p|P|$, a final contradiction. $\triangle$

Since every p-subgroup of G is contained in a maximal p-subgroup the following corollary follows immediately.

Corollary 1.2.6 *If H is a p-subgroup of G then H is contained in a Sylow p–subgroup of G.*

1.3 Coset Enumeration

Recall that a group G is *finitely presented* if it is determined by a finite set of generators which are subject to a finite set of defining relations. Typical notation is $G = \langle x_1, \ldots, x_m \mid r_1 = 1, \ldots, r_n = 1 \rangle$. The *relators* $r_1, \ldots, r_m$ are *words* of the form $y_1 y_2 \cdots y_\ell$, where each y_i is either one of the generators x_j or its inverse. The actual meaning of the notation is that $G = F_X/N$, where F_X is the free group based on the set X and N is the normal closure in F_X of the subgroup generated by the set R of relators (e.g. see [31]).

In 1936 J. A. Todd and H. S. M. Coxeter published a paper ([62]) giving a procedure for enumerating the cosets of a subgroup in a finitely presented group. The subgroup must be given by a finite set of generators, which are words from $X^\pm = X \cup X^{-1}$, X being the set of generators for the group.

In order to describe and illustrate the procedure let us suppose that G has generators $X = \{x_1, \ldots, x_m\}$ and relators $R = \{r_1, \ldots, r_n\}$, and that H has generators $S = \{s_1, \ldots, s_k\}$. Suppose further that $r_i = x_{i1} x_{i2} \cdots x_{ik_i}$ and that $s_i = y_{i1} y_{i2} \cdots y_{i\ell_i}$, all i, each x_{ij} and y_{ij} being in $X^\pm$.

The cosets of H in G are labeled by positive integers, beginning with $1 := H$. The generators s_i (as words) are each put at the top of a *subgroup table*, which will have just one row in it. At the beginning and ending of that row are the label 1 for the coset H, reflecting the fact that $1 s_i = 1$ (i.e. $H s_i = H$, since $s_i \in H$). Each relator r_i (also as a word) is put at the top of a *relator table*, which may have many rows (one for each coset, as they become labeled). At the outset the label 1 is placed at the beginning and end of the first row of each relator table, reflecting the fact that each $r_i = 1$, and hence $1 r_i = 1$ (*i.e.*, $H r_i = H$).

Two further tables are set up, basically for bookkeeping purposes – a *coset table*, in which are stored the relations between cosets, and a *definitions/bonuses (or D/B) table*, which will be explained shortly.

The procedure is started by defining a second (presumably different) coset labeled 2, and placing it in a blank position in one of the subgroup or relator tables, normally next to one of the 1's. For example, since r_1 begins with x_{11} the definition could be $2 := 1x_{11}$, meaning that the 2nd coset is Hx_{11}. Once coset 2 has been defined 2's are placed at the beginning and end of the second rows of each relator table, the definition is written in the definition part of the D/B table, and the fact that $1x_{11} = 2$ (as well as that $2x_{11}^{-1} = 1$) is recorded in the coset table. Then the subgroup and relator tables are *scanned* to see whether the information available is sufficient to fill in any further blanks. After the scanning another coset is defined, and the process continues.

When a row fills up in a subgroup or relator table it is because some coset k fills the last blank space just to the left of another coset ℓ, and at the top of the table between k and ℓ is a generator (or inverse), say y, forcing on us the information that $ky = \ell$. Three possible situations then arise: (1) It may be already known that $ky = \ell$, in which case we simply continue. (2) It may be already known that $ky = j$, where j and ℓ had been presumed different from each other. In this case they must be acknowledged to be the same (this is usually called a *collapse*), and it is necessary to replace the larger of the two by the smaller in all tables, then proceed. (3) The fact that $ky = \ell$ is simply new information (called a *bonus*, which is entered in all tables, including the bonus part of the D/B table.

When (and if) all rows of all tables fill up consistently, then the number of rows in each relator table is the index $[G\colon H]$ (for a careful proof of this fact see page 11 of [48].

Let us illustrate the procedure first with a very simple example, viz. $G = \langle x \mid x^3 = 1 \rangle$, and we take $H = 1$, so no subgroup tables will be required and the procedure will simply determine the order of G. The relator table, coset table, and D/B table begin as follows.

x	x	x
1		1

	x	x^{-1}
1		

D	B

Define $2 := 1x$ and enter the information in the tables.

x	x	x
1	2	1
2		1	2

	x	x^{-1}
1	2	
2		1

D	B
$2 = 1x$	

The 1 in the second row of the relator table resulted from scanning. Next define $3 := 2x$ and enter the information. Before scanning we have the following.

x	x	x	
1	2	3	$\rightarrow$ 1
2		1	2
3			3

	x	x^{-1}
1	2	
2	3	1
3		2

D	B
$2 = 1x$	
$3 = 2x$	

Notice that the first row of the relator table has filled up, and we have obtained a bonus, viz. $3x = 1$ (marked for convenience in the table by an arrow). When that information is entered we see that all blanks are filled in the coset table, ensuring that all rows in the relator table can be completed, and the procedure finishes.

x	x	x	
1	2	3	$\rightarrow$ 1
2	3	1	2
3	1	2	3

	x	x^{-1}
1	2	3
2	3	1
3	1	2

D	B
$2 = 1x$	$3x = 1$
$3 = 2x$	

We see that $|G| = 3$ (as expected, of course), but also that the procedure has provided a permutation representation of G, viz. the multiplication action of the generator x on the set of 3 cosets of the trivial subgroup. Directly from the coset table we see that $x \mapsto (1\,2\,3)$, determining a homomorphism from G into S_3 (in fact an isomorphism from G to A_3).

For another example, take $G = \langle x, y \mid x^3 = y^2 = (xy)^2 = 1 \rangle$, this time with $H = \langle x^2 y \rangle$. Thus we begin with a subgroup table for the generator of H (recall that it will have only one row), 3 relator tables, and as always the coset table and D/B table. After the first definition $2 := 1x$ has been made, and after scanning, the situation is as follows.

```
     x   x   y              x   x   x
1    2       1          1   2       1
                        2       1   2

     y   y                  x   y   x   y
1        1              1   2           1
2        2              2               2
```

```
      | x | y | x^-1              D    | B
1  2  |   |   |               2 = 1x   |
2     |   |   |  1
```

Note that there is no column for y^{-1} in the coset table, since it is clear from the presentation that $y^{-1} = y$. Defining $3 := 2x$ we obtain the bonuses $3y = 1$ in the subgroup table, then $3x = 1$ in the first relator table. Further scanning in the third relator table yields $2y = 2$, and all tables can be completed.

```
     x   x   y                  x   x   x
1    2   3  → 1            1     2   3  → 1
                          2         1     2
                          3               3

     y   y                      x   y   x   y
1        1                1      2 → 2   3   1
2        2                2                  2
3        3                3                  3
```

```
      | x | y | x^-1              D     | B
1  2  | 2 | 3 | 3             2 = 1x    | 3y = 1
2     | 3 | 2 | 1             3 = 2x    | 3x = 1
3     | 1 | 1 | 2                       | 2y = 2
```

Thus $[G{:}H] = 3$, and the permutation action of G on the cosets is given by $x \mapsto (1\,2\,3)$, $y \mapsto (1\,3)$, providing a homomorphism from G onto S_3. Observe, though, that $(x^2 y)^2 = 1$ (why?), so $|H| \le 2$ and $|G| = [G{:}H]|H| \le 6$, and in fact $G \cong S_3$.

Note that we can determine a (Schreier) transversal T for H in G directly from the definition table. We begin with $1 \in 1$ ($= H$), then observe that $2 = 1x$, so we may take x as a representative for 2. Next $3 = 2x = 1x^2$, so we take x^2 as a representative in 3, and we have $T = \{1, x, x^2\}$.

For a final example take $G = \langle x, y \mid xy^2 = 1, x^2y^3 = 1 \rangle$ and $H = 1$. Begin with the definitions $2 := 1x$, $3 := 2y$, and $4 := 2x$. The first bonus is $3y = 1$, then, from the first row of the 2nd relator table, $4y = 2$. When that information is inserted in the second row of the first table a further bonus of $2y = 2$ is obtained, and the tables are as follows.

	x	y	y		
1	2	3	$\rightarrow$	1	
2	4	2	$\rightarrow$	2	
3		2		3	
4				4	

	x	x	y		y	y	
1	2	4	$\rightarrow$	2	3	1
2	4					2
3					2	3
4						4

	x	y	x^{-1}	y^{-1}
1	2			3
2	4	3	1	4
3		1		2
4		2	2	

D	B
$2 = 1x$	$3y = 1$
$3 = 2y$	$4y = 2$
$4 = 2x$	$2y = 2$

Observe now, though, that we have $2y = 3$ by definition and $2y = 2$ as a bonus, so we must conclude that $3 = 2$, a *collapse*. Normally this would call for replacing each 3 by a 2, copying all the information from 3's row into 2's row, removing 3's row, and continuing. In this case, however, we see from the coset table that $2y = 3$, which becomes $2y = 2$, and $3y = 1$, which becomes $2y = 1$, so also $2 = 1$, another collapse. Finally, $1x = 2$ and $2x = 4$ allows us to conclude as well that $4 = 1$, and we have total collapse. All tables fill (with 1's), and we conclude that $|G| = 1$.

It is not difficult to draw the same conclusion from the relations: $1 = x^2y^3 = x(xy^2)y = xy$, then $1 = xy^2 = (xy)y = y$; hence also $x = y^{-1} = 1$.

The last example can be generalized by setting

$$G = \langle x, y \mid x^n y^{n+1} = x^{n+1} y^{n+2} = 1 \rangle,$$

where n is any positive integer. Just as above it is easy to check that G is trivial, but if the cosets of $H = 1$ are enumerated it is clear that $2n$ cosets must be defined before any row can fill up, thus before there is any hope of a collapse occurring. This simple example shows that there can be no general bound in terms of the index of the subgroup for the number of possibly superfluous cosets that may have to be defined to carry out the enumeration.

Exercises

1. Let $G = \langle x, y, z, u, v \mid xy = z, yz = u, zu = v, uv = x, vx = y \rangle$. Enumerate the cosets of $H = \langle x \rangle$ in G, and conclude that G is cyclic (in fact it is cyclic of order 11).

2. Do a coset enumeration to determine the order of

$$G = \langle x, y \mid xyx = y, \, yxy = x \rangle.$$

   Do you recognize G?

It is easy to tell from a completed coset table for H in G whether or not $H \triangleleft G$. Say H has generators $\{s_1, \ldots, s_k\}$ and $[G : H] = k$. Then $H \triangleleft G$ if and only if $x s_i x^{-1} \in H$ for $1 \le i \le k$ and all $x \in G$, which is if and only if $(Hx)s_i = Hx$ for all i and x. But the cosets Hx are now labeled as $1, \ldots, k$, so that condition becomes simply $\ell s_i = \ell$ for every coset ℓ and every generator s_i of H.

In the second example above, with $H = \langle x^2 y \rangle$, we observe e.g. that $2x^2y = 3xy = 1y = 3 \neq 2$, so H is not normal in G.

Coset enumeration is at the heart of several important procedures in computational group theory, including the sophisticated version of the Schreier-Sims algorithm mentioned in the previous section, a method for finding a presentation for a subgroup of a finitely presented group, and a method for determining all conjugacy classes of subgroups of low index in a finitely presented group. All of these procedures, and more, are available in GAP [57]. For a more thorough discussion of coset enumeration and some of its ramifications see [48], [43], or [44].

1.4 Semidirect Products

We review briefly the definition and construction of semidirect products of groups, as they will prove useful in later chapters.

Recall that a group G is an *extension* of K by H if $K \triangleleft G$ and $G/K \cong H$. As a special case G is called a *split extension*, or (internal) *semidirect product*, of K by H if $K \triangleleft G$, $H \le G$, $K \cap H = 1$, and $G = KH$. It will be denoted by $G = K \rtimes H$. For example if $G = S_n$, $K = A_n$, and $H = \langle (1\,2) \rangle$, then $G = K \rtimes H$.

Note that if $G = K \rtimes H$ then every $x \in G$ can be written uniquely as $x = kh$, with $k \in K$, $h \in H$, and the multiplication proceeds as follows:

$$(k_1 h_1)(k_2 h_2) = k_1 h_1 k_2 h_1^{-1} h_1 h_2 = (k_1 \cdot {}^{h_1} k_2)(h_1 h_2).$$

To construct a semidirect product (external) we require the groups K and H, together with a homomorphism $\theta: H \to \operatorname{Aut}(K)$. For present purposes it will be useful and suggestive to write the action of any $\varphi \in \operatorname{Aut}(K)$ using what might be called *left exponential* notation; i.e. if $k \in K$ then ${}^\varphi k$ is the image of k under the action of φ.

Now the construction: define G *as a set* to be the cartesian product $K \times H$, and define the product via

$$(k_1, h_1)(k_2, h_2) = (k_1 \cdot {}^{\theta h_1} k_2, h_1 h_2).$$

It will be denoted $G = K \rtimes_\theta H$, or simply by $G = K \rtimes H$ if θ is understood.

Exercise

Show that $G = K \rtimes_\theta H$ is a group and that it has subgroups $K_1 \cong K$ (normal) and $H_1 \cong H$ such that $G = K_1 \rtimes H_1$ (internal).

It is common practice to identify K with K_1 and H with H_1 (in the exercise), thereby blurring the distinction between internal and external semidirect products.

For a familiar class of examples take $K = C_n = \langle a \rangle$, cyclic of order n, $H = C_2 = \langle b \rangle$, cyclic of order 2, and define $\theta: H \to \operatorname{Aut}(K)$ via ${}^{\theta b} a = a^{-1}$. Then $D_n = H \rtimes K$ is the dihedral group of order $2n$.

For another class of examples let K be any group, set $H = \operatorname{Aut}(K)$, and take θ to be the identity map. Then the semidirect product $G = K \rtimes H$ is called the *holomorph* of K, denoted $\operatorname{Hol}(K)$.

Exercises

1. If K is Klein's 4-group then $\operatorname{Hol}(K)$ is a familiar group; determine which one.

2. If G is a group and $H \le G$ recall that H is *characteristic* in G (notation: H char G) if ${}^\varphi h \in H$ for all $\varphi \in \operatorname{Aut}(G)$ and all $h \in H$. For any group K and $L \le K$ show that L char K if and only if $L \lhd \operatorname{Hol}(K)$.

1.5   Wreath Products

Suppose that G is a group and that S is a set. Write $G^S = \{f: S \to G\}$, the set of all functions from S to G. Then G^S is a group, the operation being point-wise multiplication: $(fg)(s) = f(s)g(s)$. In fact G^S is simply the cartesian product of $|S|$ copies of G.

If G, H are groups, and if H acts on a set S, then also H acts on G^S by means of $^x f(s) = f(s^x)$, all $x \in H$, $f \in G^S$, and $s \in S$. It is easy to check that $^{(xy)}f = {}^x({}^y f)$.

The action of H on G^S determines a homomorphism θ from H to $\mathrm{Aut}(G^S)$, given by $\theta(x) = \theta_x : f \mapsto {}^x f$, all $x \in H$, $f \in G^S$. To see that observe first that $^x(fg)(s) = (fg)(s^x) = f(s^x)g(s^x) = {}^x f(s) \cdot {}^x g(s) = ({}^x f \cdot {}^x g)(s)$, so θ_x is an endomorphism of G^S. Next, $\ker \theta_x$ consists of all those $f \in G^S$ for which $^x f(s) = f(s^x) = 1 \in G$ for all $s \in S$, hence for all $s^x \in S$, i.e. $f = 1 \in G^S$. Each θ_x is onto, since $\theta_x(^{x^{-1}}f) = f$ for any $f \in G^S$, so $\theta_x \in \mathrm{Aut}(G^S)$. Finally, check that $\theta_{xy} = \theta_x \theta_y$, all x, $y \in H$.

We may now define the *wreath product* of G and H, relative to S, denoted $G \wr H$, to be the semidirect product $G^S \rtimes_\theta H$. Note that $|G \wr H| = |G|^{|S|}|H|$. The multiplication is given by $(f, x)(g, y) = (f \cdot {}^x g, xy)$.

Exercise

Show that $G \wr H$ is abelian if and only if both G and H are abelian and the action of H on S is trivial.

For an easy example take $G = H = \mathrm{Sym}(2)$, with the natural action of H on $S = \{1, 2\}$. Then $G \wr H$ has a normal subgroup isomorphic with $G \times G$, a Klein 4-group, and $|G \wr H| = 8$, so $G \wr H$ is dihedral.

Suppose that G, H, and S are as above, and suppose further that G acts on a set T. We may then define a group action of $G \wr H$ on the cartesian product $T \times S$ via $(t, s)^{(f, x)} = (t^{f(s)}, s^x)$. It is straightforward to check that the action really is a group action.

Proposition 1.5.1 *If G acts on T with kernel K and H acts on S with kernel L then the action of $G \wr H$ on $T \times S$ has kernel $K \wr L$.*

Proof We have (f, x) in the kernel of the action if and only if $(t, s)^{(f, x)} = (t^{f(s)}, s^x) = (t, s)$ for all $t \in T$, $s \in S$, or equivalently $s^x = s$ and $t^{f(s)} = t$, all s and t, i.e. $x \in L$ and $f(s) \in K$, all s, or $f \in K^S$, as required. $\triangle$

Corollary 1.5.2 *The wreath product $G \wr H$ acts faithfully on $T \times S$ if and only if G and H act faithfully on T and S, respectively.*

More general notions of wreath product exist; e.g. see [60]. In general, wreath products are often used for constructing pathological examples. For example a wreath product construction can be used to construct an infinite p-group having trivial center and a proper subgroup that is its own normalizer. For present purposes the construction will mainly be used to describe Sylow subgroups of symmetric groups.

Exercise

If $(f, x) \in G \wr H$ calculate $(f, x)^{-1}$ explicitly.

1.6   Transitivity and Primitivity

Suppose that a group G acts on a set S, with $|S| = n$, finite. If $k \leq n$ say that G acts k-*transitively* if, for each pair of k-tuples $(a_1, \ldots, a_k)$ and $(b_1, \ldots, b_k)$ of *distinct* elements of S, there is some $x \in G$ for which $a_i^x = b_i$, $1 \leq i \leq k$. Clearly k-transitivity implies ℓ-transitivity for $1 \leq \ell \leq k$. For an obvious example note that the symmetric group S_n acts n-transitively on $\{1, \ldots, n\}$.

Exercise

Show that a transitive group G of degree n is k-transitive, for $2 \leq k \leq n$, if and only if the stabilizer G_s is $(k-1)$-transitive on $S \setminus \{s\}$ for each $s \in S$.

Proposition 1.6.1 *If $n \geq 3$ then the alternating group A_n acts $(n-2)$-transitively, but not $(n-1)$-transitively, on $\{1, \ldots, n\}$.*

Proof We use induction on n, beginning with the fact that A_3 is transitive, but not 2-transitive, on $\{1, 2, 3\}$. For $n > 3$ we have $\text{Stab}_{A_n}(n) = A_{n-1}$, and A_{n-1} is $(n-3)$-transitive on $\{1, \ldots, n-1\}$ by the induction hypothesis, so A_n is $(n-2)$-transitive by the exercise above. If it were $(n-1)$-transitive there would be some $\sigma \in A_n$ fixing each of 1, 2, $\ldots$, $n-2$ and carrying $n-1$ to n. But the only $\sigma \in S_n$ that does this is the transposition $(n-1 \ \ n) \notin A_n$. $\triangle$

It is customary to view S_n and A_n as *trivial* examples of highly transitive groups. Every group has a transitive action (on itself – that is Cayley's theorem). We shall see later many examples of doubly and triply transitive groups, and even some 4- and 5-transitive groups. It is a very deep fact that there are no nontrivial k-transitive groups for $k \geq 6$.

If G is transitive on S define the *rank* of G to be the number of G_s-orbits in S (for any $s \in S$). Clearly the rank is at least 2 (if $|S| \geq 2$), since $\{s\}$ is a G_s-orbit. Note that if G is k-transitive on S and $k \geq 2$ then $\text{rank}(G) = 2$.

Proposition 1.6.2 *The rank of a transitive permutation group G on a set S is equal to the number of distinct double cosets $G_s x G_s$ of the stabilizer G_s for any $s \in S$.*

Proof Fix $s \in S$. For $x, y \in G$, s^x and s^y lie in the same G_s-orbit if and only if $s^{x G_s} = s^{y G_s} = s^{G_s x G_s} = s^{G_s y G_s}$. The latter equality holds if and only if $G_s x G_s = G_s y G_s$, for if $s^{G_s x G_s} = s^{G_s y G_s}$ then there are $a, b \in G_s$ such that $s^{axb} = s^y$, hence $axby^{-1} \in G_s$, $axb \in G_s y \subseteq G_s y G_s$, and so $G_s x G_s \subseteq G_s y G_s$. The reverse inclusion is similar, and they are equal. $\triangle$

Corollary 1.6.3 *If G is k-transitive on S, $k \geq 2$, then*

$$G = G_s \cup G_s x G_s$$

for any $s \in S$ and any $x \in G \setminus G_s$.

Proposition 1.6.4 *If G is k-transitive on S and $|S| = n$ then*

$$\frac{n!}{(n-k)!} \,\Big|\, |G|.$$

In fact, if $T \subseteq S$ and $|T| = k$ then $[G:G_T] = n!/(n-k)!$.

Proof Write $T = \{s_1, \ldots, s_k\}$. Then $|G| = n \cdot |G_{s_1}|$ by Proposition 1.2.1, and G_{s_1} is $(k-1)$-transitive on $S \setminus \{s_1\}$ by the exercise on page 16. Similarly $|G_{s_1}| = (n-1)|G_{s_1,s_2}|$, so $|G| = n(n-1)|G_{s_1,s_2}|$, etc. $\triangle$

Say that G is *sharply k-transitive* on S if it is k-transitive and there is a subset $T \subseteq S$, $|T| = k$, with $G_T = 1$. If G is finite and $|S| = n$, finite, then the definition is equivalent, by Proposition 1.6.4, with requiring that $|G| = n!/(n-k)!$. Thus, for example, S_n is sharply n-transitive, and A_n is sharply $(n-2)$-transitive, since $n!/(n-(n-2))! = n!/2 = |A_n|$.

We shall see examples of sharply 2-transitive and sharply 3-transitive groups in Chapter 2.

For examples of sharply 1-transitive actions think back again to Cayley's theorem. The *right regular* action of any group G on itself is determined by right multiplication: $x : y \mapsto yx$; it is transitive and the stabilizer of any point is trivial.

In general, if G is sharply 1-transitive on S we say that G is *regular* on S, in which case $|S| = |G|$. Note, in fact, that if we fix $s \in S$ then $S = \{s^x : x \in G\}$, and if $x, y \in G$ then $(s^x)^y = s^{xy}$. It follows easily that the action of G on S is equivalent with the right regular action of G on itself.

We shall return to regular actions after introducing the notion of primitivity.

If G acts on S then a *block* is a subset $B \subseteq S$ such that $B \neq S$, $|B| > 1$, and if $x \in G$ then either $B^x = B$ or $B \cap B^x = \emptyset$. If G is transitive on S and has *no* blocks we say that G is *primitive* on S.

For a negative example let G be the dihedral group D_4 acting on the vertices $\{1, 2, 3, 4\}$ of a square. If 1 and 3 are opposite vertices then clearly $\{1, 3\}$ is a block, so D_4 is not primitive.

Exercise

If G is transitive but imprimitive on S and B is a block for G show that $|B| \big| |S|$. Conclude that if $|S| = p$, a prime, then G, if transitive, must be primitive.

Proposition 1.6.5 *Suppose that G acts transitively on S. Then G is primitive if and only if each stabilizer G_s, $s \in S$, is a maximal (proper) subgroup of G.*

Proof Exercise. $\triangle$

Proposition 1.6.6 *If G is doubly transitive on S then G is primitive.*

Proof Suppose that $B \subseteq S$, with $|B| > 1$ and $B \neq S$. Choose $a, b \in B$, $a \neq b$, and $c \in S \setminus B$. Then choose $x \in G$ with $a^x = a$ and $b^x = c$. Then $a \in B^x \cap B$ so $B^x \cap B \neq \emptyset$, and $c \in B^x \setminus B$, so $B^x \neq B$. Thus B is not a block for G.   △

Proposition 1.6.7 *Suppose that G is primitive on S and $K \triangleleft G$ is the kernel of the action. Suppose that $N \triangleleft G$ and $N \not\leq K$. Then N is transitive on S.*

Proof Since $N \not\leq K$ there is an N-orbit in S having more than one element, say $B = \mathrm{Orb}_N(a)$ and $|B| \geq 2$. If $x \in G$ then $B^x = a^{Nx} = a^{xN}$ (since $N \triangleleft G$). Thus $B^x = \mathrm{Orb}_N(a^x)$. Since S is partioned into its N-orbits we have either $B^x = B$ or $B^x \cap B = \emptyset$, and consequently $B = S$, or else B would be a block for G.   △

Proposition 1.6.8 *Suppose that G acts on S, $N \leq G$, and N is transitive on S. If $s \in S$ then $G = \mathrm{Stab}_G(s) \cdot N$.*

Proof If $x \in G$ then $s^x = s^y$ for some $y \in N$. Thus $s^{xy^{-1}} = s$, so $xy^{-1} \in \mathrm{Stab}_G(s)$ and $x \in \mathrm{Stab}_G(s) \cdot y \subseteq \mathrm{Stab}_G(s) \cdot N$.   △

The next theorem will be applied in Chapter 2.

Theorem 1.6.9 (Iwasawa) *Suppose that G is faithful and primitive on S and $G' = G$. Fix $s \in S$ and set $H = \mathrm{Stab}_G(s)$. Suppose there is a solvable subgroup $K \triangleleft H$ such that $G = \langle \cup \{K^x : x \in G\} \rangle$. Then G is simple.*

Proof Suppose that $1 \neq N \triangleleft G$. By Proposition 1.6.7 N is transitive on S, and by Proposition 1.6.8 $G = HN$. Thus $KN \triangleleft HN = G$. If $x \in G$ then $K^x \leq (KN)^x = KN$, so $\cup \{K^x : x \in G\} \subseteq KN$ and hence $KN = G$. Since K is solvable $K^{(m)} = 1$ for some $K^{(m)}$ in the derived series. Check inductively that $(KN)^{(\ell)} \leq K^{(\ell)}N$ for all ℓ. Thus $G = G^{(m)} = (KN)^{(m)} \leq K^{(m)}N = N$ and $N = G$.   △

Theorem 1.6.10 *Suppose that G is primitive and faithful on S, and that G_s is simple, $s \in S$. Then either*

  1. G is simple, or

  2. G has a normal subgroup N that is regular on S.

Proof If G is not simple choose $N \triangleleft G$, $1 \neq N \neq G$. If $s \in S$ then $N \cap G_s \triangleleft G_s$, which is simple, so $N \cap G_s = G_s$ or 1. If $N \cap G_s = G_s$ then $G_s \leq N$, but $G_s \neq N$ since N is transitive by Proposition 1.6.7. But then $N = G$ by Proposition 1.6.5, a contradiction. Thus $N \cap G_s = 1$, i.e. $N_s = 1$, all $s \in S$, so N is regular on S.   △

Theorem 1.6.10 could be a useful simplicity criterion if regular normal subgroups were relatively rare, so let us consider them.

Note first that if $N \lhd G$ then G acts by conjugation on $N^\star = N \setminus \{1\}$. If G also acts on a set S, then each G_s, $s \in S$, acts on $N^\star$ as well.

Proposition 1.6.11 *Suppose that G acts on S, $N \lhd G$, and N is regular on S. If $s \in S$ then the action of G_s on $N^\star$ by conjugation is equivalent with the action of G_s on $S \setminus \{s\}$.*

Proof Since N is regular on S we have $|N| = |S|$ and $S = \{s^y : y \in N\}$. Thus $f(y) = s^y$ defines a bijection from $N^\star$ to $S \setminus \{s\}$. If $x \in G_s$ then $f(y^x) = s^{x^{-1}yx} = s^{yx} = f(y)^x$, so the actions are equivalent. $\triangle$

Recall that a finite group is an *elementary abelian p-group* (p a prime) if it is isomorphic with a direct product of cyclic groups of order p.

Theorem 1.6.12 *Suppose G acts on S, $N \lhd G$, and N is regular on S.*

1. *If G is doubly transitive on S then N is elementary abelian and $|S| = p^k$ for some prime p and some $k \in \mathbb{N}$.*

2. *If G is 3-transitive on S then either N is an elementary abelian 2-group or N is cyclic of order 3, so $|S| = 2^k$ or 3.*

3. *If G is 4-transitive on S then N is a Klein four-group and $|S| = 4$.*

4. *G cannot be 5-transitive.*

Proof (1) Since G is doubly transitive G_s is transitive on $S \setminus \{s\}$, and therefore also on $N^\star$ by Proposition 1.6.11. Choose a prime p such that $p \big| |N|$ and choose $x \in N$ with $|x| = p$. By transitivity $|y| = p$ for all $y \in N^\star$, so N is a p-group. The action (by conjugation) of each $x \in G_s$ on N determines an automorphism of N, so $Z(N)^x = Z(N)$, and $Z(N) \neq 1$, hence $Z(N) = N$ by transitivity; i.e. N is abelian. Thus $N \cong \mathbb{Z}_p^k$ for some k by the structure theorem for finite abelian groups.

(2) Now G_s is doubly transitive on $N^\star$. By part (1) N is an elementary abelian p-group. If $p > 2$ choose $x \in N^\star$ and note that $x \neq x^{-1}$. The stabilizer in G_s of x is transitive on $N^\star \setminus \{x\}$, but also any $z \in \mathrm{Stab}_{G_s}(x)$ effects an automorphism of N by conjugation, so $(x^{-1})^z = x^{-1}$. It follows that $N^\star \setminus \{x\} = \{x^{-1}\}$, hence $N = \{1, x, x^{-1}\} \cong \mathbb{Z}_3$.

(3) Now G_s is 3-transitive on $N^\star$, so $|N^\star| \geq 3$ and $|N| \geq 4$. By parts (1) and (2) N is an elementary abelian 2-group. If $|N| \geq 8$ then N contains a Klein four-group K as a subgroup, say with $K^\star = \{a, b, c\}$, and also some $d \in N \setminus K$. By 3-transitivity there is some $x \in G_s$ with $a^x = a$, $b^x = b$, and $c^x = d$. But then $d = c^x = (ab)^x = a^x b^x = ab = c$, a contradiction, so in fact $|N| = 4$.

(4) If G_s were 4-transitive then we would have $|N^\star| \geq 4$, but also, by part (3), $|N| = 4$, a contradiction. $\triangle$

As an application we may (re-)prove the simplicity of the alternating groups A_n for $n \geq 5$.

To begin, recall that A_5 has conjugacy classes of sizes 1, 12, 12, 15, and 20, so A_5 is simple by Lagrange's theorem. This starts an induction. Assume that $n \geq 6$ and that A_{n-1} is simple. Write $G = A_n$ and $G_n = \text{Stab}_G(n) = A_{n-1}$. Then G is 4-transitive but of degree greater than 4, so it cannot have a regular normal subgroup by Theorem 1.6.12; hence it is simple by Theorem 1.6.10.

1.7   Some Linear Algebra

The material developed in this section will be used in the next chapter in the discussion of symplectic groups.

Suppose that F is a field and V is a vector space over F. Then its *dual space*, $V^\star = \text{Hom}_F(V)$, is the set of all *linear functionals* from V to F; it is also an F-vector space. If $\{v_1, \ldots, v_n\}$ is a basis for V define $\{v_1^\star, \ldots, v_n^\star\} \subseteq V^\star$ via $v_i^\star(v_j) = \delta_{ij}$, all i and j (and of course extend by linearity). This defines the *dual basis* for $V^\star$. If $\sum_i a_i v_i^\star = 0$ then $0 = \sum_i a_i v_i^\star(v_j) = a_j$, all j; if $f \in V^\star$ set $f(v_i) = b_i$ and check that $f = \sum_i b_i v_i^\star$. In particular $dim_F(V^\star) = dim_F(V)$ (when finite).

A *bilinear form* on V is a function $B: V \times V \to F$ that becomes a linear functional in either variable if the other variable is fixed, i.e.

$$B(u + v, w) = B(u, w) + B(v, w) \text{ and } B(au, w) = aB(u, w);$$
$$B(u, v + w) = B(u, v) + B(u, w) \text{ and } B(u, aw) = aB(u, w)$$

for all $u, v, w \in V$, all $a \in F$.

If B is a bilinear form on V and $\{v_1, \ldots, v_n\}$ is a basis for V set $b_{ij} = B(v_i, v_j)$, all i, j. Then $\widehat{B} = [b_{ij}]$ is called the matrix of B relative to $\{v_i\}$. If it becomes necessary to exhibit the dependence on the basis we will write $\widehat{B}_{\{v_i\}}$. If $u, w \in V$ write $u = \sum_i a_i v_i$ and $w = \sum_i c_i v_i$, so u and w are represented by column vectors (coordinate vectors) $\widehat{u} = (a_1, \ldots, a_n)^t$ and $\widehat{w} = (c_1, \ldots, c_n)^t$. Then

$$B(v, w) = B\left(\sum_i a_i v_i, \sum_j c_j v_j\right) = \sum_{i,j} a_i B(v_i, v_j) c_j = \widehat{v}^t \widehat{B} \widehat{w}.$$

Conversely, if $\widehat{B} = [b_{ij}]$ is any $n \times n$ matrix over F we may define a bilinear form (depending on a choice $\{v_i, \ldots, v_n\}$ of basis) by decreeing that $B(v_i, v_j) = b_{ij}$ and extending by linearity.

If $\{w_1, \ldots, w_n\}$ is another basis write $w_j = \sum_i d_{ij} v_i$, $d_{ij} \in F$, for all j. Then $B(w_i, w_j) = \sum_{k,l} d_{ki} B(v_k, v_l) d_{l,j} = \sum_{k,l} d_{ki} b_{kl} d_{l,j}$, the ij-entry of $D^t \widehat{B} D$, where $D = [d_{ij}]$ is the (invertible) change-of-basis matrix. In particular, any two representing matrices have the same rank, which we define to be the rank of the form B.

Classically, square matrices M and N are called *congruent* if $M = D^t N D$ for some invertible matrix D, so two representing matrices for a bilinear form B are congruent. Note that then $\det M = (\det D)^2 \det N$.

Write $F^{\times 2} = \{a^2 : a \in F^\star\}$, the subgroup of squares in the multiplicative group $F^\star$. If B is a bilinear form on V and $\widehat{B}$ is a representing matrix define the *discriminant* of B to be

$$\text{discr}(B) = \begin{cases} 0 & \text{if } \det \widehat{B} = 0, \\ (\det \widehat{B}) F^{\times 2} \in F^\star / F^{\times 2} & \text{otherwise.} \end{cases}$$

Note that $\text{discr}(B)$ is independent of the choice of basis. Say that B is *nondegenerate* if $\text{discr}(B) \neq 0$.

If B is a bilinear form on V define two maps L and R from V to $V^\star$ as follows: $L: v \mapsto L_v$, $R: v \mapsto R_v$, with $L_v(w) = B(v, w)$ and $R_v(w) = B(w, v)$, all $w \in V$. It is easy to check that L and R are linear transformations.

Define the *left* and *right radicals* of V (relative to B) to be

$$\text{rad}_L(V) = \ker L = \{v \in V : B(v, w) = 0, \text{ all } w \in V\},$$

$$\text{rad}_R(V) = \ker R = \{v \in V : B(w, v) = 0, \text{ all } w \in V\}.$$

Proposition 1.7.1 *A bilinear form B on V is nondegenerate if and only if $\text{rad}_L(V) = 0$ if and only if $\text{rad}_R(V) = 0$.*

Proof Choose a basis $\{v_1, \dots, v_n\}$ and write $\widehat{B} = [b_{ij}]$ for the matrix representing B. Note that $v \in \text{rad}_R(V)$ if and only if $B(v_i, v) = 0$, all i. Write $v = \sum_j a_j v_j$, $a_j \in F$. Then $B(v_i, v) = \sum_j b_{ij} a_j = 0$, all i, if and only if the column vector $(a_1, \dots, a_n)^t$ is a solution to the homogeneous system $\widehat{B} X = 0$. Thus $\text{rad}_R(V) = 0$ if and only if $\det \widehat{B} \neq 0$, and the equivalence with $\text{rad}_L(V) = 0$ is similar. $\triangle$

Corollary 1.7.2 *A bilinear form B on V is nondegenerate if and only if $V^\star = \text{Im } L = \text{Im } R$, i.e. if and only if given any $f \in V^\star$ there exist $u, v \in V$ such that $f(w) = B(w, u) = B(v, w)$, all $w \in V$.*

The corollary generalizes.

Corollary 1.7.3 *Suppose that B is a nondegenerate bilinear form on V and that W is a subspace of V. If $f \in W^\star$ then there exist $u, v \in V$ such that $f = R_u|_W = L_v|_W$.*

Proof We may choose a basis $\{w_1, \dots, w_k\}$ for W and enlarge it to a basis $\{w_1, \dots, w_n\}$ for V. Extend f to $f_1 \in V^\star$ via $f_1|_W = f$, $f_1(v_i) = 0$ for $i > k$. By the first corollary $\exists u, v \in V$ with $f_1 = R_u = L_v$, and hence $f = R_u|_W = L_v|_W$. $\triangle$

For any subset $S \subseteq V$ define

$$\perp_L (S) = \{v \in V : B(v,w) = 0, \text{ all } w \in S\},$$

$$\perp_R (S) = \{v \in V : B(w,v) = 0, \text{ all } w \in S\}.$$

Thus $v \in \perp_L (S)$ if and only if $L_v|_S = 0$, $v \in \perp_R (S)$ if and only if $R_v|_S = 0$. Note that $\perp_L (V) = \text{rad}_L(V)$ and $\perp_R (V) = \text{rad}_R(V)$.

Exercise

Show that

1. $\perp_L (S)$ and $\perp_R (S)$ are subspaces of V,
2. $\perp_L (\perp_R (S)) \supseteq S$ and $\perp_R (\perp_L (S)) \supseteq S$, and
3. if $S \subseteq T$ then $\perp_L (T) \subseteq \perp_L (S)$ and $\perp_R (T) \subseteq \perp_R (S)$.

Proposition 1.7.4 *If B is a nondegenerate bilinear form on V and W is a subspace of V then*

$$\dim \perp_L (W) = \dim \perp_R (W) = \dim V - \dim W.$$

Proof The map $\theta : v \mapsto L_v|_W$ is a linear transformation from V onto $W^\star$ (Corollary 1.7.3), and $\ker \theta = \perp_L (W)$ by definition. Thus $\dim V = \dim W^\star + \dim \perp_L (W)$, and similarly $\dim V = \dim W^\star + \dim \perp_R (W)$. $\triangle$

Corollary 1.7.5 *If B is a nondegenerate bilinear form on V and W is a subspace of V then $\perp_L (\perp_R (W)) = \perp_R (\perp_L (W)) = W$.*

Proof By the exercise above $\perp_L (\perp_R (W)) \supseteq W$, and

$$
\begin{aligned}
\dim \perp_L (\perp_R (W)) &= \dim V - \dim \perp_R (W) \\
&= \dim V - (\dim V - \dim W) = \dim W.
\end{aligned}
$$

$\triangle$

Exercise

If B is nondegenerate on V and S is any subset of V show that

$$\perp_L (\perp_R (S)) = \perp_R (\perp_L (S))$$

is the subspace spanned by S.

Suppose that B is a bilinear form on V. Say that B is *symmetric* if $B(v,w) = B(w,v)$ for all $v, w \in V$. Say that B is *alternate* if $B(v,v) = 0$ for all $v \in V$. Note that B is symmetric if and only if any representing matrix $\widehat{B}$ is symmetric, i.e. $\widehat{B}^t = \widehat{B}$.

If B is an alternate form and $v, w \in V$ then $0 = B(v + w, v + w) = B(v, v) + B(v, w) + B(w, v) + B(w, w)$, so $B(w, v) = -B(v, w)$. If char $F = 2$ this shows that B is symmetric; in general it shows that B is *skew symmetric* (also called *antisymmetric*). If char $F \neq 2$ then the concepts of alternate form and skew symmetric form are equivalent, for then skew symmetry entails $B(v, v) = -B(v, v)$, or $B(v, v) = 0$. In general, a representing matrix $\widehat{B} = [b_{ij}]$ for an alternate form is an *alternate*, or *skew symmetric*, matrix, meaning that $b_{ii} = 0$, all i, and $\widehat{B}^t = -\widehat{B}$.

Proposition 1.7.6 *Suppose that B is a bilinear form on V satisfying*

$$B(u, v)B(w, u) = B(v, u)B(u, w) \qquad (*)$$

for all $u, v, w \in V$. Then B is symmetric or alternate.

Proof Take $u = v$ and conclude that

$$B(v, v)[B(w, v) - B(v, w)] = 0 \qquad (**)$$

for all $v, w \in V$. We wish to show that either $B(v, v) = 0$ for *all* $v \in V$, or $B(w, v) = B(v, w)$ for *all* $v, w \in V$. If that is not the case then $\exists x, y, z \in V$ with $B(y, y) \neq 0$ and $B(x, z) \neq B(z, x)$. Apply $(**)$ and obtain (i) $B(x, x) = B(z, z) = 0$, (ii) $B(x, y) = B(y, x)$, and (iii) $B(y, z) = B(z, y)$. Use $(*)$, with $u = x$, $v = y$, and $w = z$, to conclude from (ii) that $B(x, y) = B(y, x) = 0$. Interchange u and w in $(*)$ and get, again with $u = x$, $v = y$, and $w = z$, that $B(z, y) = B(y, z) = 0$ from (iii). Thus $B(x, y + z) = B(x, z) \neq B(z, x) = B(y + z, x)$, and, by $(**)$,

$$0 = B(y + z, y + z) = B(y, y) + B(y, z) + B(z, y) + B(z, w) = B(y, y) \neq 0,$$

a contradiction. $\triangle$

For any bilinear form B on V, and $v, w \in V$, say that v is *orthogonal* to w, and write $v \perp w$, if $B(v, w) = 0$. If orthogonality is a reflexive relation on V (i.e. $v \perp w \Rightarrow w \perp v$) then the form B is called *reflexive*. Note that if B is a reflexive form then $\perp_L (S) = \perp_R (S)$ for all subsets $S \subseteq V$; we agree to write $S^\perp = \perp_L (S) = \perp_R (S)$ in that case. If W is a subspace then $W^\perp$ is called the *orthogonal complement* of W. Note that $W \cap W^\perp = 0$ if and only if $B|_{W \times W}$ is a nondegenerate form on W. If $W \cap W^\perp = 0$ we say that W is a *nondegenerate subspace* (relative to B).

In general, if B is a reflexive bilinear form on V and W is a subspace we define the *radical* of W to be rad $W = W \cap W^\perp$.

Exercise

Suppose that B is a reflexive bilinear form on V, and $V = U \oplus W$, with $U \perp W$. Show that rad $V =$ rad $U \oplus$ rad W.

Proposition 1.7.7 *A bilinear form B on V is reflexive if and only if it is either symmetric or alternate.*

Proof $\Leftarrow$: A symmetric form is clearly reflexive. If B is alternate it is skew symmetric, so $v \perp w$ means $0 = B(v, w) = -B(w, v)$, and therefore $w \perp v$.

$\Rightarrow$: Take any $u, v, w \in V$ and set $x = B(u, v)w - B(u, w)v$. Then $B(u, x) = B(u, v)B(u, w) - B(u, w)B(u, v) = 0$, so $u \perp x$. Thus also $x \perp u$, which says $0 = B(x, u) = B(u, v)B(w, u) - B(v, u)B(u, w)$. Apply Proposition 1.7.6. $\triangle$

Suppose that B_1, B_2 are bilinear forms on spaces V_1, V_2, respectively. An *isometry* relative to B_1 and B_2 is an F-isomorphism $\sigma: V_1 \to V_2$ satisfying $B_2(\sigma v, \sigma w) = B_1(v, w)$ for all $v, w \in V_1$. If an isometry exists the forms are called *equivalent*.

Proposition 1.7.8 *Bilinear forms B_1, B_2 on spaces V_1, V_2 are equivalent if and only if there are bases for V_1, V_2 relative to which $\widehat{B_1} = \widehat{B_2}$.*

Proof $\Rightarrow$: Let $\sigma: V_1 \to V_2$ be an isometry. Choose a basis $\{v_1, \ldots, v_n\}$ for V_1, and set $w_i = \sigma v_i$, all i, to get a basis for V_2. Then $\widehat{B_1} = [B_1(v_i, v_j)] = [B_2(\sigma v_i, \sigma v_j)] = [B_2(w_i, w_j)] = \widehat{B_2}$.

$\Leftarrow$: Suppose $\widehat{B_1} = \widehat{B_2}$ relative to bases $\{v_i\}$ and $\{w_i\}$ for V_1 and V_2. Define $\sigma: V_1 \to V_2$ via $\sigma(v_i) = w_i$, all i, an isomorphism. If $u, w \in V_1$ write $u = \sum_i a_i v_i$ and $w = \sum_i b_i v_i$. Then

$$
\begin{aligned}
B_1(u, w) &= \sum_{i,j} a_i b_j B_1(v_i, v_j) = \sum_{i,j} a_i b_j B_2(w_i, w_j) \\
&= \sum_{i,j} a_i b_j B_2(\sigma v_i, \sigma v_j) = B_2(\sum_i a_i \sigma v_i, \sum_j b_j \sigma v_j) = B_2(\sigma u, \sigma w).
\end{aligned}
$$

$\triangle$

Proposition 1.7.9 *Suppose that B is a reflexive bilinear form on V, and that W is a nondegenerate subspace of V. Then $V = W \oplus W^\perp$.*

Proof We may choose a basis $\{v_1, \ldots, v_k\}$ for W and adjoin further vectors $\{v_{k+1}, \ldots, v_n\}$ to extend it to a basis for V. Represent B by its matrix $\widehat{B} = [b_{ij}]$ relative to $\{v_i\}$, and represent vectors in V correspondingly as column vectors in F^n. If $v = \sum_i a_i v_i \in V$, so $v \mapsto (a_1, \ldots, a_n)^t$, then $v \in W^\perp$ if and only if $0 = B(v_i, v) = \sum_j B(v_i, v_j)a_j = \sum_j b_{ij}a_j$ for $1 \le i \le k$, i.e. if and only if $(a_1, \ldots, a_n)^t$ is in the solution space of the homogeneous system $AX = 0$, whose coefficient matrix A is the $k \times n$ matrix over F whose rows are the first k rows of $\widehat{B}$. Thus dim $W^\perp = n - \text{rank } A \ge n - k$. But W nondegenerate means that rad $W = W \cap W^\perp = 0$, so $W + W^\perp = W \oplus W^\perp$, and $\dim(W \oplus W^\perp) = \dim W + \dim W^\perp \ge k + (n-k) = n$; hence $W \oplus W^\perp = V$.
$\triangle$

Assume for the remainder of this section that B is an alternate bilinear form on V.

For a simple example take V of dimension 2 and $\widehat{B} = \begin{bmatrix} 0 & 1 \\ -1 & 0 \end{bmatrix}$. Thus, if $u = \begin{bmatrix} x_1 \\ y_1 \end{bmatrix}$ and $v = \begin{bmatrix} x_2 \\ y_2 \end{bmatrix}$, then $B(u,v) = x_1 y_2 - x_2 y_1$. We'll see that this example is surprisingly general.

If $u, v \in V$ and $B(u,v) \neq 0$ then $\{u,v\}$ is linearly independent, for $B(u, au) = aB(u,u) = 0$. If $B(u,v) = b \neq 0$ set $u_1 = b^{-1}u$ and $v_1 = v$. Then $B(u_1, v_1) = 1$. The subspace W spanned by u_1 and v_1 is called a *hyperbolic plane*, with *hyperbolic pair* (u_1, v_1) as basis. Note that the restriction of B to W has representing matrix $\begin{bmatrix} 0 & 1 \\ -1 & 0 \end{bmatrix}$ relative to a hyperbolic pair as basis.

A bit of notation: If U and W are subspaces of V, with $U \cap W = 0$ and also $U \perp W$, we will denote their sum (which is a direct sum) by $U \oplus V$.

Theorem 1.7.10 *If B is an alternating form on V then*

$$V = W_1 \oplus W_2 \oplus \cdots \oplus W_r \oplus \operatorname{rad} V,$$

a direct sum of mutually orthogonal subspaces, with each W_i a hyperbolic plane. Consequently V has a basis

$$\{u_1, v_1, \ldots, u_r, v_r, w_1, \ldots, w_{n-2r}\}$$

relative to which the representing matrix $\widehat{B}$ has block diagonal form

$$\begin{bmatrix} M & & & 0 \\ & \ddots & & \\ & & M & \\ 0 & & & 0_{n-2r} \end{bmatrix}, \quad where \ M = \begin{bmatrix} 0 & 1 \\ -1 & 0 \end{bmatrix}.$$

Proof If $B = 0$ then $\operatorname{rad} V = V$ and there is nothing to prove. Otherwise, choose a hyperbolic pair (u_1, v_1), as above, and let W_1 be the hyperbolic plane that they span. Then W_1 is a nondegenerate subspace (the determinant of the restriction of B to W_1 is 1), so $V = W \oplus W_1^\perp$ by Proposition 1.7.9. Also $\operatorname{rad} V = V^\perp = (W_1 \oplus W_1^\perp)^\perp = W_1^\perp \cap W_1^{\perp\perp} = \operatorname{rad}(W_1^\perp)$. The result follows by induction on the dimension. $\triangle$

Corollary 1.7.11 *B has even rank; if B is nondegenerate then $\dim V$ is even.*

Corollary 1.7.12 *Alternate bilinear forms B_1 and B_2 on spaces V_1 and V_2 are equivalent if and only if $\dim V_1 = \dim V_2$ and $\operatorname{rank} B_1 = \operatorname{rank} B_2$.*

Proof Proposition 1.7.8.   △

Corollary 1.7.13 *The determinant of any representing matrix for an alternating form is a square in F.*

Proof If $\widehat{B}$ is as in the theorem then its determinant is 0 or 1; relative to another basis the matrix takes the form $D^t \widehat{B} D$, with determinant $(\det D)^2 \det \widehat{B}$.   △

Chapter 2

Some Groups

The point of this chapter is simply the introduction of several examples, and classes of examples, of groups, in fairly complete detail. They will serve as a useful source of illustrative examples in later chapters.

2.1 $\mathrm{Aut}(\mathbb{Z}_n)$

We write $\mathbb{Z}_n$ for the ring $\mathbb{Z}/n\mathbb{Z}$ of integers mod n. Throughout this section we will abuse the notation as is customary and let context determine whether an integer k is being viewed as an integer or is taken mod n and hence is being viewed as an element of $\mathbb{Z}_n$.

If $\theta \in \mathrm{End}(\mathbb{Z}_n)$ say that $\theta(1) = m$. Then for any $k \in \mathbb{Z}_n$ we have $\theta(k) = \theta(1 + \cdots + 1) = k\theta(1) = m \cdot k$, and it follows easily (verify!) that $\mathrm{End}(\mathbb{Z}_n) \cong \mathbb{Z}_n$. Also, $\theta \in \mathrm{End}(\mathbb{Z}_n)$ is an automorphism if and only if $\theta(1) = m$ is another generator of $\mathbb{Z}_n$, i.e. if and only if $(m, n) = 1$, so $\mathrm{Aut}(\mathbb{Z}_n) \cong U(\mathbb{Z}_n)$, the group of units in $\mathbb{Z}_n$.

If $n = \prod_{i=1}^{k} p_i^{e_i}$ is the prime factorization of n then by the Chinese Remainder Theorem $\mathbb{Z}_n \cong \bigoplus_i \mathbb{Z}_{p_i^{e_i}}$, so

$$\mathrm{Aut}(\mathbb{Z}_n) \cong U(\mathbb{Z}_n) \cong U(\mathbb{Z}_{p_1^{e_1}}) \times \cdots \times U(\mathbb{Z}_{p_k^{e_k}}).$$

Thus we may concentrate on $\mathrm{Aut}(\mathbb{Z}_{p^e})$, p a prime. Recall from elementary algebra that

$$|\mathrm{Aut}(\mathbb{Z}_{p^e})| = \varphi(p^e) = p^{e-1}(p-1),$$

where φ denotes Euler's totient function.

Proposition 2.1.1 *If p is an odd prime then* $\mathrm{Aut}(\mathbb{Z}_{p^e})$ *is cyclic.*

Proof If $e = 1$ then $\mathbb{Z}_p$ is a field and it is a standard fact that $U(\mathbb{Z}_p) = \mathbb{Z}_p^\star$ is cyclic. Suppose, then, that $e > 1$. Choose a generator x for $U(\mathbb{Z}_p)$, and note that $y = x + p$ is also a generator for $U(\mathbb{Z}_p)$. If $x^{p-1} = 1$ in $\mathbb{Z}_{p^2}$

27

then $y^{p-1} = (x+p)^{p-1} = x^{p-1} + (p-1)px^{p-2} = 1 - px^{p-2} \neq 1$ in $\mathbb{Z}_{p^2}$
since $p \nmid x$. Thus we may assume that $x^{p-1} = 1$ in $\mathbb{Z}_p$ but that $x^{p-1} \neq 1$
in $\mathbb{Z}_{p^2}$. Let us show that the multiplicative order of x in $\mathbb{Z}_{p^e}$ is $\varphi(p^e) = p^{e-1}(p-1)$. As above we may write $x^{p-1} = 1 + pa$, with $p \nmid a$. Thus
$x^{\varphi(p^e)} = (x^{p-1})^{p^{e-1}} = (1+pa)^{p^{e-1}} = 1 + \sum_{k=1}^{p^{e-1}} \binom{p^{e-1}}{k}(pa)^k = 1$ in $\mathbb{Z}_{p^e}$, so
$|x^{p-1}| \mid p^{e-1}$. However, $(x^{p-1})^{p^{e-2}} = (1+pa)^{p^{e-2}} = 1 + ap^{e-1} \neq 1$ in $\mathbb{Z}_{p^e}$,
since $p \nmid a$. Thus $|x^{p-1}| = p^{e-1}$. Suppose now that $x^m = 1$ in $\mathbb{Z}_{p^e}$. Then also
$(x^{p-1})^m = 1$, so $p^{e-1} \mid m$. Write $m = p^{e-1}\ell$. Then $1 = x^m = (x^{p^{e-1}})^\ell = x^\ell$ in
$\mathbb{Z}_p$ (since $x^p = x$ in $\mathbb{Z}_p$), so $p-1 \mid \ell$. Thus $p^{e-1}(p-1) \mid m$, and we conclude
that $|x| = |U(\mathbb{Z}_{p^e})|$.   △

It should be remarked that the proof above is not constructive. Even in
the case of $\mathbb{Z}_p$ there seems to be no efficient algorithm for finding a generator
– called in number theory a *primitive root*.

Note that $U(\mathbb{Z}_2) = 1$ and $U(\mathbb{Z}_4) = \{1,3\} = \langle 3 \rangle$. However, $U(\mathbb{Z}_8) = \{1,3,5,7\} = \langle -1 \rangle \times \langle 5 \rangle$.

Proposition 2.1.2 *If $e \geq 3$ then* $\mathrm{Aut}(\mathbb{Z}_{2^e}) \cong U(\mathbb{Z}_{2^e}) = \langle -1 \rangle \times \langle 5 \rangle$.

Proof Use induction on e to see that $5^{2^{e-3}} = 1 + 2^{e-1}$ in $\mathbb{Z}_{2^e}$. It is true for
$e = 3$, so assume the result for some $e \geq 3$ and consider $e+1$. Note that
$(1 + 2^{e-1})^2 = 1 + 2^e + 2^{2e-2}$ in $\mathbb{Z}$, and that $2e - 2 \geq e + 1$. We have assumed
that $5^{2^{e-3}} = 1 + 2^{e-1}$ in $\mathbb{Z}_{2^e}$, i.e. that $5^{2^{e-3}} + b \cdot 2^e = 1 + 2^{e-1}$ in $\mathbb{Z}$ for some
$b \in \mathbb{Z}$. Squaring, we have

$$5^{2^{e-2}} + b \cdot 5^{2^{e-3}} 2^{e+1} + b^2 2^{2e} = 1 + 2^e + 2^{2e-2}. \qquad (*)$$

Since $2e - 2 \geq e + 1$ this says that $5^{2^{e-2}} = 5^{2^{(e+1)-3}} = 1 + 2^{(e+1)-1}$ in
$\mathbb{Z}_{2^{e+1}}$. Thus $|5| > 2^{e-3}$ in $\mathbb{Z}_{2^e}$. But also $(*)$ shows that $5^{2^{e-2}} = 1$ in $\mathbb{Z}_{2^e}$, so
$|5| = 2^{e-2}$.

To finish the proof it will be sufficient to show that the elements $(-1)^i 5^j$,
for $0 \leq i \leq 1$ and $0 \leq j < 2^{e-2}$, are all distinct. Say $(-1)^i 5^j = (-1)^k 5^\ell$ in
$\mathbb{Z}_{2^e}$, with the exponents in the indicated ranges. Then $(-1)^i = (-1)^k$ in $\mathbb{Z}_4$,
and it follows that $i = k$. But then also $j = \ell$, since $|5| = 2^{e-2}$.   △

We summarize in a theorem.

Theorem 2.1.3 *Suppose that $n \in \mathbb{N}$ has prime factorization $n = 2^e \prod_{i=1}^k p_i^{e_i}$.
Then*

$$\mathrm{Aut}(\mathbb{Z}_n) \cong U(\mathbb{Z}_n) \cong U(\mathbb{Z}_{2^e}) \times U(\mathbb{Z}_{p_1^{e_1}}) \times \cdots \times U(\mathbb{Z}_{p_k^{e_k}});$$

$U(\mathbb{Z}_2)$, $U(\mathbb{Z}_4)$, *and* $U(\mathbb{Z}_{p_i^{e_i}})$ *are all cyclic; if $e \geq 3$ then* $U(\mathbb{Z}_{2^e}) = \langle -1 \rangle \times \langle 5 \rangle$.

Corollary 2.1.4 $\mathrm{Aut}(\mathbb{Z}_n)$ *is abelian; it is cyclic precisely when $n = 2$, 4, p^k,
or $2p^\ell$, p an odd prime.*

Proof Exercise.   △

2.2 Metacyclic Groups

A group G is *metacyclic* if it is an extension of a cyclic group by a cyclic group, i.e. if G has a normal cyclic subgroup $A = \langle a \rangle$ with $G/A = \langle Ab \rangle$ also cyclic. If the extension splits over A, so there is a cyclic subgroup $B = \langle b \rangle \leq G$ with $G = A \rtimes B$, then G is called *split metacyclic*.

Suppose that G is metacyclic, as above, and finite, say with $|A| = |a| = m$ and $[G : A] = s$, so $|G| = ms$. Then $b^s \in A$, so there is some $t \in \mathbb{Z}, 0 \leq t < m$, with $b^s = a^t$. Note that if $t = 0$ then G splits. Conversely, if G splits over A then it is possible to choose b so that $t = 0$ (however, see Exercise 2 on page 30).

Since $A \lhd G$ there is also $r \in \mathbb{Z}, 1 \leq r < m$, with $b^{-1}ab = a^r$; note that $(r, m) = 1$, since $|a^r| = |a|$. For all $k \geq 1$ we have $b^{-k}ab^k = a^{r^k}$, and in particular $a = b^{-s}ab^s = a^{r^s}$, so $r^s \equiv 1 \pmod{m}$. Note also that $a^{t(r-1)} = (a^r)^t a^{-t} = (b^{-1}ab)^t a^{-t} = b^{-1}a^t b a^{-t} = b^{-1}b^s b a^{-t} = a^t a^{-t} = 1$, so $tr \equiv t \pmod{m}$.

For convenience we gather all this arithmetic information in the next proposition.

Proposition 2.2.1 *If $G = \langle a, b \rangle$ is metacyclic with parameters m, s, t, and r as above, then*

$$(r, m) = 1, \quad r^s \equiv 1 \pmod{m}, \quad and \quad tr \equiv t \pmod{m}.$$

Since $ab = ba^r$ all elements of G can be written in the form $b^i a^j$, and in fact

$$G = \{ b^i a^j : 0 \leq i \leq s - 1, \ 0 \leq j \leq m - 1 \}.$$

Proposition 2.2.2 *Suppose that G is split metacyclic, with $t = 0$, and assume that $(m, s) = 1$. Then all Sylow subgroups of G are cyclic.*

Proof Since $|G| = ms$ and $(m, s) = 1$ it is clear that Sylow subgroups of $A = \langle a \rangle$ and of $B = \langle b \rangle$ are Sylow in G, and all Sylow subgroups of G (up to conjugacy) appear in either A or B. $\triangle$

The converse of Proposition 2.2.2 is true but not quite so simple; it will be proved in Chapter 4 (Theorem 4.1.17).

The final proposition of this section provides a presentation for a finite metacyclic group.

Proposition 2.2.3 *Suppose that $m, s, t, r \in \mathbb{N}$ satisfy $1 \leq r < m$, $0 \leq t < s$, $(r, m) = 1$, $r^s \equiv 1 \pmod{m}$, and $tr \equiv t \pmod{m}$, and set*

$$G = \langle a, b \ | \ a^m = 1, b^s = a^t, b^{-1}ab = a^r \rangle.$$

Then G is metacyclic of order ms, with $A = \langle a \rangle \lhd G$ and $G/A = \langle Ab \rangle$.

Exercises

1. Prove Proposition 2.2.3. (Suggestion: construct such a group from cyclic groups A and B of orders m and s by defining an appropriate multiplication on the *set* $A \times B$.)

2. If $G = \langle a, c \mid a^8 = 1, c^2 = a^6, c^{-1}ac = a^5 \rangle$ show that G is split metacyclic, even though $t \neq 0$. (*Hint* Try replacing c by $b = ca^i$ for some appropriate choice of i.)

3. If $G = \langle a, b \mid a^4 = b^3 = 1, b^{-1}ab = a^2 \rangle$ show that G is cyclic of order 6.

4. If $G = \langle a, b \mid a^4 = 1, b^4 = a^3, b^{-1}ab = a^3 \rangle$ show that G is cyclic of order 2.

5. If G is metacyclic, as in Proposition 2.2.1, show that its derived group is $G' = \langle a^{r-1} \rangle$, of order $m/(m, r-1)$. Determine the center $Z(G)$.

Exercises 3 and 4 are intended to point out the necessity of the arithmetic conditions in Proposition 2.2.3.

Probably the best known specific examples of groups as in Proposition 2.2.3 are the dihedral groups (which are split) of order $2m$,

$$D_m = \langle a, b \mid a^m = b^2 = 1, b^{-1}ab = a^{m-1} \rangle,$$

and the generalized quaternion groups (which are not split) of order $4m$,

$$Q_m = \langle c, d \mid c^{2m} = 1, d^2 = c^m, d^{-1}cd = c^{2m-1} \rangle.$$

2.3   Sylow Subgroups of Symmetric Groups

If $p \in \mathbb{Z}$ is a prime and $0 \neq m \in \mathbb{Z}$ we write $\nu_p(m)$ to denote the exact power of p that divides m, *i.e.* $p^{\nu_p(m)} \mid m$ but $p^{\nu_p(m)+1} \nmid m$. Also, if $a \in \mathbb{R}$ we write $\lfloor a \rfloor$ to denote the greatest integer that is less than or equal to a. Examples: $\nu_5(1000) = 3 = \lfloor \pi \rfloor$.

If $p \in \mathbb{N}$ is a prime set $J_p = \langle (1\,2\,\cdots\,p) \rangle$, a cyclic subgroup of order p in the symmetric group S_p. Set $S = \{1, 2, \ldots, p\}$. By Corollary 1.5.2 the wreath product $J_p \wr J_p$ is a faithful permutation group on $S \times S$; it has order $p^p \cdot p = p^{p+1}$ and degree p^2. Thus $J_p \wr J_p$ is isomorphic with a subgroup of S_{p^2}.

Since

$$|S_{p^2}| = (p^2)! = (p \cdot p) \cdots ((p-1)p) \cdots ((p-2)p) \cdots \cdots (2p) \cdots p \cdots 1$$

we have $\nu_p((p^2)!) = p + 1$, so in fact $J_p \wr J_p$ is isomorphic with a Sylow p-subgroup of S_{p^2}.

Our goal is to generalize from p^2 to arbitrary $n > 0$ in $\mathbb{N}$.

Proposition 2.3.1 *Suppose that a, b, c, $\in \mathbb{Z}$, with b, $c > 0$. Then*

$$\lfloor \lfloor a/b \rfloor /c \rfloor = \lfloor a/bc \rfloor.$$

Proof Write $a = bq + r$, with $0 \leq r < b$, and $q = cq_1 + r_1$, with $0 \leq r_1 < c$. Then $\lfloor a/b \rfloor = q$ and $\lfloor q/c \rfloor = \lfloor \lfloor a/b \rfloor /c \rfloor = q_1$. But also $a = b(cq_1 + r_1) + r = bcq_1 + br_1 + r$, and $0 \leq br_1 + r < b(c-1) + b = bc$, so $a/bc = q_1 + (br_1 + b)/bc$, hence also $\lfloor a/bc \rfloor = q_1$.   △

Proposition 2.3.2 *If p is a prime and $0 < n \in \mathbb{Z}$ then*

$$\nu_p(n!) = \lfloor n/p \rfloor + \lfloor n/p^2 \rfloor + \cdots.$$

Proof Note that $n! = 1 \cdots p \cdots (2p) \cdots \cdots (\lfloor n/p \rfloor p) \cdots n$, and

$$\prod_{k=1}^{\lfloor n/p \rfloor} (kp) = p^{\lfloor n/p \rfloor} \cdot \lfloor n/p \rfloor!.$$

Thus $\nu_p(n!) = \lfloor n/p \rfloor + \nu_p(\lfloor n/p \rfloor!)$. The result now follows by induction and Proposition 2.3.1.   △

For an example take $p = 7$ and $n = 1000$. Then $\lfloor 1000/7 \rfloor = 142$, $\lfloor 1000/7^2 \rfloor = \lfloor 142/7 \rfloor = 20$, and $\lfloor 1000/7^3 \rfloor = \lfloor 20/7 \rfloor = 2$, so $\nu_7(1000!) = 164$.

Given arbitrary p and n, as above, write $n = a_0 + a_1 p + \cdots + a_k p^k$, with $a_k \neq 0$, $a_i \in \mathbb{Z}$, and $0 \leq a_i < p$, all i. Then

$$\lfloor n/p \rfloor = a_1 + a_2 p + \cdots + a_k p^{k-1},$$

$$\lfloor n/p^2 \rfloor = a_2 + a_3 p + \cdots + a_k p^{k-2},$$

etc., and so

$$\nu_p(n!) = a_1 + a_2(1+p) + a_3(1+p+p^2) + \cdots + a_k(1+p+\cdots+p^{k-1})$$

$$= \sum_{j=1}^{k} a_j \sum_{i=0}^{j-1} p^i = \sum_{j=1}^{k} a_j \left(\frac{p^j - 1}{p - 1} \right)$$

is the order of a Sylow p-subgroup of S_n.

Back to wreath products.

Recall that $J_p = \langle (1\,2 \cdots p) \rangle \leq S_p$. Define $J_p^{wr(0)} = 1$, $J_p^{wr(1)} = J_p$, $J_p^{wr(2)} = J_p \wr J_p$, and inductively $J_p^{wr(r)} = J_p \wr J_p^{wr(r-1)}$ for $r \geq 3$. We shall call $J_p^{wr(r)}$ the rth *wreath power* of J_p.

Proposition 2.3.3 *If $1 \leq r \in \mathbb{Z}$ then $J_p^{wr(r)}$ has order $p^{1+p+\cdots+p^{r-1}}$ and degree p^r. It is isomorphic to a subgroup of S_{p^r}.*

Proof By Corollary 1.5.2, and induction, $J_p^{wr(r)}$ is faithful on $S^r = S \times S \times \cdots \times S$, so its degree is p^r. Assume inductively that $|J_p^{wr(r-1)}| = p^{1+p+\cdots+p^{r-2}}$. Then

$$|J_p^{wr(r)}| = |J_p \wr J_p^{wr(r-1)}| = p^{p^{r-1}} \cdot p^{1+p+\cdots+p^{r-2}} = p^{1+p+\cdots+p^{r-1}}.$$

$\triangle$

If G and H are groups acting on sets S and T, respectively, then $G \times H$ acts naturally on $S \times T$ via $(s,t)^{(g,h)} = (s^g, t^h)$. In particular $G^2 = G \times G$ acts on $S^2 = S \times S$, and likewise G^m acts on S^m if $1 \le m \in \mathbb{N}$; if G has degree d then G^m has degree d^m.

Theorem 2.3.4 *If* $1 \le n \in \mathbb{N}$ *and* $p \in \mathbb{N}$ *is prime write* $n = a_0 + a_1 p + \cdots + a_k p^k$, *with* $a_k \ne 0$, $a_i \in \mathbb{Z}$, *and* $0 \le a_i < p$, *all* i. *Then* $P = \prod_{i=0}^{k}(J_p^{wr(i)})^{a_i}$ *is isomorphic with a Sylow p-subgroup of the symmetric group* S_n.

Proof Note that the group P has order $\prod_{i=1}^{k} |J_p^{wr(i)}|^{a_i} = p^{\nu_p(n!)}$ and degree $\sum_{i=0}^{k} a_i p^i = n$. The action is faithful so P is isomorphic with a Sylow p-subgroup of S_n. $\triangle$

For example, if $n = 1000$ and $p = 7$ we have seen that $\nu_7(1000!) = 164$, and $1000 = 6 + 2 \cdot 7 + 6 \cdot 7^2 + 2 \cdot 7^3$. Thus a Sylow 7-subgroup of S_{1000} is isomorphic with

$$(J_7^{wr(0)})^6 \times (J_7^{wr(1)})^2 \times (J_7^{wr(2)})^6 \times (J_7^{wr(3)})^2.$$

2.4 Affine Groups of Fields

If F is a field we write $F^\star$ for its multiplicative group $F \setminus \{0\}$. Denote by $\text{Aff}(F)$ the group (under composition) of all functions $\tau_{b,a} \colon F \to F$, where $a \in F^\star$, $b \in F$, and $\tau_{b,a}(x) = ax + b$, all $x \in F$. Thus if F is finite, say $|F| = q$, then $|\text{Aff}(F)| = q(q-1)$.

Note that $\tau_{b,a}\tau_{d,c} = \tau_{b+ad,ac}$, so $\text{Aff}(F) \cong F \rtimes_\theta F^\star$, where θ maps the multiplicative group $F^\star$ to automorphisms of the additive group F via $\theta(a) \colon d \mapsto ad$. Internally $\text{Aff}(F)$ has the normal subgroup $T = \{\tau_{b,1} \colon b \in F\}$ of translations, and the subgroup $H = \{\tau_{0,a} \colon a \in F^\star\}$, with $\text{Aff}(F) = T \rtimes H$.

Clearly $\text{Aff}(F) \le \text{Perm}(F)$, so we may naturally view $\text{Aff}(F)$ as a permutation group acting on F.

Proposition 2.4.1 *If* F *is a field then the affine group* $G = \text{Aff}(F)$ *is sharply 2-transitive on* F.

Proof Clearly the translation subgroup T is transitive on F, so G is transitive. Furthermore, the subgroup $H = \{\tau_{0,a} \colon a \in F^\star\}$ is $\text{Stab}_G(0)$, and H is transitive on $F^\star$, so G is doubly transitive on F by the exercise on page 16. Finally, if $\tau = \tau_{b,a} \in G$ stabilizes both 0 and 1 then $0 = \tau(0) = b$ and $1 = \tau(1) = a$, so $\tau = 1$ and G is sharply 2-transitive. $\triangle$

Corollary 2.4.2 *If p is a prime and $0 < n \in \mathbb{N}$ then there exists a sharply 2-transitive group of degree p^n.*

Exercises

Let $G = \mathrm{Aff}(F) = T \rtimes H$ as above.

1. If $\tau \in G \setminus H$ show that $H \cap H^\tau = 1$. Conclude that $H = N_G(H)$.

2. If $1 \neq \tau \in T$ show that $C_G(\tau) = T$.

The results in the two exercises above reflect the fact that $\mathrm{Aff}(F)$ is an example of a *Frobenius* group, of which we shall hear more later.

2.5 Finite Groups in 2 and 3 Dimensions

If V is a real vector space with an inner product then the *orthogonal group* $O(V)$ is the group of all linear transformations τ on V that preserve the inner product, *i.e.* $(\tau u, \tau v) = (u, v)$ for all $u, v \in V$. Equivalently $\tau \in O(V)$ if and only if τ preserves *lengths* of vectors.

Recall from linear algebra that if $\tau \in O(V)$ then $\det \tau = \pm 1$, and that if $\lambda \in \mathbb{C}$ is an eigenvalue of τ then $|\lambda| = 1$. If A is the matrix representing τ relative to an orthonormal basis for V then A is an *orthogonal* matrix, meaning that its transpose is its inverse.

In this section we shall take V to be either $\mathbb{R}^2$ or $\mathbb{R}^3$, in each case with the standard inner product, commonly called the *dot product*. We will write $e_1, e_2, \ldots$ for the standard orthonormal basis vectors in each case. Vectors will be column vectors, e.g. $v = \begin{bmatrix} a \\ b \end{bmatrix}$, but for typographical convenience they will often be written as row vectors, $v = (a, b)$.

Suppose first that $\tau \in O(\mathbb{R}^2)$. Say that $\tau(e_1) = (a, b)$. Then $a^2 + b^2 = 1$. Choose θ, $0 \le \theta < 2\pi$, so that $a = \cos\theta$, $b = \sin\theta$. We have $\tau e_2 \perp \tau e_1$, so $\tau e_2 = \pm(b, -a)$.

Case 1. If $\tau e_2 = (-b, a)$ then τ is represented by the matrix

$$A = \begin{bmatrix} \cos\theta & -\sin\theta \\ \sin\theta & \cos\theta \end{bmatrix},$$

$\det A = 1$, and τ is a (counterclockwise) *rotation* through angle θ about the origin as center.

Case 2. If $\tau e_2 = (b, -a)$ then

$$A = \begin{bmatrix} \cos\theta & \sin\theta \\ \sin\theta & -\cos\theta \end{bmatrix},$$

and det $A = -1$. If we set

$$v_1 = (\cos\frac{\theta}{2}, \sin\frac{\theta}{2}), \quad v_2 = (-\sin\frac{\theta}{2}, \cos\frac{\theta}{2})$$

it is easy to check that $v_1 \perp v_2$, $\tau v_1 = v_1$, and $\tau v_2 = -v_2$. It follows that τ is geometrically a *reflection*, with mirror the line spanned by v_1.

Note that products and inverses of rotations are rotations, and that reflections have order 2. If G is any subgroup of $O(\mathbb{R}^2)$ then the set H of rotations in G is a normal subgroup of G.

Exercise

Show that the product of two reflections in $O(\mathbb{R}^2)$ is a rotation. What is the angle of the rotation?

Suppose that G is a finite subgroup of $O(\mathbb{R}^2)$, and let H be its rotation subgroup. If $H \neq 1$ choose $\rho \neq 1$ in H whose angle $\theta = \theta(\rho)$ is minimal. For any $\tau \in H$ choose $m \in \mathbb{N}$ so that $m\theta(\rho) \leq \theta(\tau) < (m+1)\theta(\rho)$. Then $0 \leq \theta(\tau) - m\theta(\rho) < \theta(\rho)$. But $\theta(\tau) - m\theta(\rho) = \theta(\rho^{-m}\tau)$, and $\theta(\rho)$ is minimal, so $\theta(\rho^{-m}\tau) = 0$, i.e. $\rho^{-m}\tau = 1$, or $\tau = \rho^m$ and $H = \langle\rho\rangle$ is cyclic. Its order is $n = 2\pi/\theta(\rho)$; we write $H = C_n^{(2)}$.

If $G \neq H$ choose a reflection $\sigma \in G \setminus H$. Since $\det(\rho^k\sigma) = -1$ for each k the coset $H\sigma$ contains n distinct reflections. If $\tau \in G$ is any reflection then $\det(\tau\sigma) = 1$, so $\tau\sigma \in H$ and $\tau \in H\sigma$. Thus H has index 2 in G, and

$$G = \langle\rho, \sigma\rangle = \{1, \rho, \rho^2, \ldots, \rho^{n-1}, \sigma, \rho\sigma, \ldots, \rho^{n-1}\sigma\}$$

of order $2n$. Since $\sigma\rho$ is a reflection we have $(\sigma\rho)^2 = 1$, or $\sigma\rho = \rho^{-1}\sigma$, so G is dihedral. It will be denoted by $\mathcal{D}_n^{(2)}$.

We have proved the next theorem, which has been attributed by Hermann Weyl [65] to Leonardo da Vinci.

Theorem 2.5.1 *If G is a finite subgroup of $O(\mathbb{R}^2)$ then G is either a cyclic group $C_n^{(2)}$ or a dihedral group $\mathcal{D}_n^{(2)}$, $n = 1, 2, 3, \ldots$.*

Next consider three dimensions: $V = \mathbb{R}^3$. If $\rho \in O(\mathbb{R}^3)$ and $\det \rho = 1$, then ρ is called a *rotation*.

Theorem 2.5.2 (Euler) *If $\rho \in O(\mathbb{R}^3)$ is a rotation then it is in fact a rotation about a fixed line as its axis, in the sense that ρ has an eigenvector v having eigenvalue 1 such that the restriction of ρ to the plane $\mathcal{P} = v^\perp$ is a rotation of $\mathcal{P}$.*

Proof If λ_1, λ_2, and λ_3 are the eigenvalues of ρ then one of them, say λ_1, must be real, as they are the roots of a cubic with real coefficients. If $\lambda_2 \notin \mathbb{R}$ then its conjugate $\overline{\lambda_2}$ is also a root, so $\overline{\lambda_2} = \lambda_3$ in that case. Since $\det \rho = \lambda_1\lambda_2\lambda_3 = 1$ the only possibilities are (relabeling if necessary)

1. $\lambda_1 = 1$, $\lambda_2 = \lambda_3 = \pm 1$, and

2. $\lambda_1 = 1$, $\lambda_2 = \overline{\lambda_3} \notin \mathbb{R}$.

In either case 1 is an eigenvalue. Choose a corresponding eigenvector v and note that also $v = \rho^{-1}\rho v = \rho^{-1}v$. If $u \perp v$ then $(\rho u, v) = (u, \rho^{-1}v) = (u, v) = 0$, so $\mathcal{P} = v^\perp$ is ρ-invariant. Since the determinant of the restriction $\rho|_{\mathcal{P}}$ is the product of the other two eigenvalues λ_2 and λ_3, we see that $\det(\rho|_{\mathcal{P}}) = 1$ and $\rho|_{\mathcal{P}}$ is a rotation of $\mathcal{P}$. $\triangle$

A *reflection* in $O(\mathbb{R}^3)$ is a transformation σ whose effect is to map every point of $\mathbb{R}^3$ to its mirror image with respect to some plane $\mathcal{P}$ through the origin. Thus it is characterized by the fact that $\sigma u = u$ for all $u \in \mathcal{P}$ and $\sigma v = -v$ if $v \in \mathcal{P}^\perp$. If r is chosen to be a unit vector in $\mathcal{P}^\perp$ (which is a line), then σ is given by the formula $\sigma v = v - 2(v, r)r$, all $v \in \mathbb{R}^3$. It is not difficult to show, by arguments similar to those in the proof of Euler's Theorem above, that if $\tau \in O(\mathbb{R}^3)$ and $\det \tau = -1$ then the effect of τ is that of a reflection through a plane $\mathcal{P}$ followed by a rotation about an axis perpendicular to $\mathcal{P}$.

If $\mathcal{P}$ is a plane (through the origin) in $\mathbb{R}^3$ then each rotation ρ in $\mathcal{O}(\mathcal{P})$ extends naturally to a rotation in $O(\mathbb{R}^3)$; it suffices to set $\rho(v) = v$ for $v \in \mathcal{P}^\perp$ and extend by linearity.

By extending each rotation in a cyclic subgroup $C_n^{(2)}$ in that way we obtain an isomorphic group of rotations in $O(\mathbb{R}^3)$, which will be denoted $C_n^{(3)}$.

On the other hand, if $\sigma \in \mathcal{O}(\mathcal{P})$ is a reflection then σ can also be extended to a *rotation* in $O(\mathbb{R}^3)$; in this case set $\sigma(v) = -v$ for $v \in \mathcal{P}^\perp$. The result is a rotation through angle π with the original reflecting line in $\mathcal{P}$ as its axis.

Thus each τ in a dihedral group $\mathcal{D}_n^{(2)}$ can be extended to a rotation in $O(\mathbb{R}^3)$, the result being an isomorphic subgroup denoted by $\mathcal{D}_n^{(3)}$.

Note that $C_2^{(3)}$ and $\mathcal{D}_1^{(3)}$ each consist of the identity and a rotation through angle π, so they are indistinguishable.

Consider next the rotational symmetry groups of regular solids in $\mathbb{R}^3$. There are (up to similarity) just five regular solids: the tetrahedron, cube, octahedron, dodecahedron, and icosahedron. Only three groups result, however, for simple geometric reasons. If midpoints of adjacent faces of a cube are joined by line segments then those line segments are the edges of an octahedron having the same rotational symmetry group as the cube. Similar remarks apply to the dodecahedron and icosahedron.

Denote by $\mathcal{T}$ the rotation group of a regular tetrahedron. Clearly $\mathcal{T}$ is transitive on the set of 4 faces, and the stabilizer of any face is cyclic of order 3, so $|\mathcal{T}| = 12$ by Proposition 1.2.1. Being isomorphic to a subgroup of Sym(4), it is clear that $\mathcal{T}$ must be isomorphic with Alt(4), the only subgroup of order 12 in Sym(4).

Denote by $\mathcal{O}$ the rotation group of a regular octahedron (or cube); $\mathcal{O}$ is transitive on the set of 6 faces of the cube and the stabilizer of a face is cyclic of order 4, so $|\mathcal{O}| = 24$. A moment's thought convinces one that $\mathcal{O}$ acts faithfully on the set of 4 diagonals of the cube, so $\mathcal{O} \cong \text{Sym}(4)$.

Note that a regular tetrahedron can be inscribed in a cube, its 6 edges being appropriately chosen diagonals of the faces of the cube. It follows that we may view $\mathcal{T}$ as a subgroup of $\mathcal{O}$.

Finally, denote by $\mathcal{I}$ the rotation group of a regular icosahedron (or dodecahedron); it is transitive on the 20 faces of the icosahedron and the stabilizer of a face has order 3, so $|\mathcal{I}| = 60$. It is possible to inscribe a cube in a dodecahedron; each of the 12 edges of the cube is an appropriately chosen "diagonal" of one of the 12 faces of the dodecahedron. In fact, since each pentagonal face has 5 different diagonals, it follows that there are exactly 5 such cubes inscribed in the dodecahedron. The group $\mathcal{I}$ acts faithfully on the set of 5 inscribed cubes, and hence $\mathcal{I} \cong \text{Alt}(5)$.

The unit sphere $\{v \in \mathbb{R}^3 : \|v\| = 1\}$ is invariant under $O(\mathbb{R}^3)$. If $1 \neq \rho \in O(\mathbb{R}^3)$ is a rotation then by Euler's Theorem (2.5.2) there are exactly two points on the unit sphere satisfying $\rho v = v$, viz. the two points where the axis of ρ pierce the sphere. Those two points will be called the *poles* of ρ.

If $G \leq O(\mathbb{R}^3)$ write $\mathcal{S} = \mathcal{S}_G$ for the set of poles of nonidentity rotations in G.

Proposition 2.5.3 *If $G \leq O(\mathbb{R}^3)$ then G acts on its set $\mathcal{S}$ of poles.*

Proof If $v \in \mathcal{S}$ then v is a pole for some rotation $\rho \in G$. If $\tau \in G$ then $(\tau \rho \tau^{-1}) \tau v = \tau \rho v = \tau v$, so τv is a pole for the rotation $\tau \rho \tau^{-1}$ and $\tau v \in \mathcal{S}$. $\triangle$

Each cyclic group $\mathcal{C}_n^{(3)}$, $n > 1$, has just two poles. No rotation in $\mathcal{C}_n^{(3)}$ carries one to the other, so $\mathcal{S}$ has two 1-point orbits, and each stabilizer is all of $\mathcal{C}_n^{(3)}$.

The dihedral group $\mathcal{D}_n^{(3)}$ has n rotation axes in the plane $\mathcal{P}$ on which $\mathcal{D}_n^{(2)}$ acts, and one more axis $\mathcal{P}^\perp$, so $\mathcal{D}_n^{(3)}$ has $2n + 2$ poles. The two poles on $\mathcal{P}^\perp$ constitute one orbit. If n is odd the n poles that are vertices of a regular n-gon in $\mathcal{P}$ make up another orbit, and the set of their negatives make up another. If n is even the vertices of the regular n-gon again make up an orbit, and the poles that are on axes joining midpoints of opposite sides of the n-gon make up a third. In both cases there are three orbits in $\mathcal{S}$, and stabilizers have orders n, 2, and 2, respectively.

It is easy to analyze the actions of $\mathcal{T}$, $\mathcal{O}$, and $\mathcal{I}$ on their sets of poles; all the data are tabulated below.

G	$|G|$	Orbits	$|\mathcal{S}|$	Orders	of	stabilizers
$\mathcal{C}_n^{(3)}$	n	2	2	n	n	
$\mathcal{D}_n^{(3)}$	$2n$	3	$2n + 2$	2	2	n
$\mathcal{T}$	12	3	14	2	3	3
$\mathcal{O}$	24	3	26	2	3	4
$\mathcal{I}$	60	3	62	2	3	5

Theorem 2.5.4 *If $G \leq O(\mathbb{R}^3)$ is a finite group of rotations then G is one of $C_n^{(3)}$, $n \geq 1$; $\mathcal{D}_n^{(3)}$, $n \geq 2$; $\mathcal{T}$; $\mathcal{O}$; or $\mathcal{I}$.*

Proof We may assume $G \neq C_1^{(3)}$. Let $\mathcal{S}$ be the set of poles of G and let $\mathcal{U}$ denote the set of all ordered pairs (ρ, v), where $1 \neq \rho \in G$ and $v \in \mathcal{S}$ is a pole of ρ. Set $|G| = n$, and for each $v \in \mathcal{S}$ set $m_v = |\text{Orb}_G(v)|$ and $n_v = |\text{Stab}_G(v)|$. Thus $n = m_v n_v$, all $v \in \mathcal{S}$.

Since each $\rho \neq 1$ in G has two poles we have $|\mathcal{U}| = 2(n-1)$. On the other hand, we may count the elements of $\mathcal{U}$ by counting the number of group elements corresponding to each pole. Suppose that $v_1, \ldots, v_k$ is a set of poles, one from each of the G-orbits in $\mathcal{S}$, and write $m_{v_i} = m_i$, $n_{v_i} = n_i$. Since each $v \in \text{Orb}(v_i)$ has $n_v = n_i$ we have

$$
\begin{aligned}
|\mathcal{U}| &= \sum_k \{n_v - 1 : v \in \mathcal{S}\} \\
&= \sum_{i=1}^k m_i(n_i - 1) = \sum_{i=1}^k (n - m_i).
\end{aligned}
$$

Thus $2n - 2 = \sum_i (n - m_i)$, and, dividing by n, we have $2 - 2/n = \sum_{i=1}^k (1 - 1/n_i)$. But $1 \leq 2 - 2/n < 2$, and each $n_i \geq 2$, so $1/2 \leq 1 - 1/n_i < 1$, and k must be either 2 or 3.

If $k = 2$ then

$$
2 - \frac{2}{n} = (1 - \frac{1}{n_1}) + (1 - \frac{1}{n_2}),
$$

or $2 = n/n_1 + n/n_2 = m_1 + m_2$, so $m_1 = m_2 = 1$, $n_1 = n_2 = n$. Thus G has just one axis of rotation and is a cyclic group $C_n^{(3)}$.

If $k = 3$ we may assume that $n_1 \leq n_2 \leq n_3$. If n_1 were 3 or greater we would have $\sum_i (1 - 1/n_i) \geq \sum_i (1 - 1/3) = 2$, a contradiction. Thus $n_1 = 2$ and

$$
2 - \frac{2}{n} = \frac{1}{2} + (1 - \frac{1}{n_2}) + (1 - \frac{1}{n_3}),
$$

or

$$
\frac{1}{2} + \frac{2}{n} = \frac{1}{n_2} + \frac{1}{n_3} > \frac{1}{2}.
$$

If n_2 were 4 or greater we would have $1/n_2 + 1/n_3 \leq 1/2$, a contradiction, so $n_2 = 2$ or 3.

If $n_2 = 2$ then $n_3 = n/2$, and $m_1 = m_2 = n/2$, $m_3 = 2$. Setting $m = n/2$ we have the data in the table above for the dihedral group $\mathcal{D}_m^{(3)}$.

If $n_2 = 3$ we have $1/6 + 2/n = 1/n_3$, and the only possibilities are (1) $n_3 = 3$, $n = 12$, (2) $n_3 = 4, n = 24$, and (3) $n_3 = 5, n = 60$; for $n_3 \geq 6$ would require $2/n < 0$. Thus in the cases 1, 2, 3 we have precisely the data for $\mathcal{T}$, $\mathcal{O}$, and $\mathcal{I}$.

In each case the group G not only shares the data in the table with one of the groups in the table, but in fact *is* the group. For example, in case 1 the poles in either of the 4-element orbits are the vertices of a regular tetrahedron centered at the origin. The tetrahedron is invariant under G so $G \leq \mathcal{T}$, but also $|G| = |\mathcal{T}|$, so $G = \mathcal{T}$. Further discussion of this point for the other groups may be found e.g. in [69]. $\triangle$

The group $\mathcal{O}^{\star} \leq O(\mathbb{R}^3)$ of *all* symmetries of a cube is larger than $\mathcal{O}$, e.g. since $-1 \in \mathcal{O}^{\star} \setminus \mathcal{O}$. Observe, however, that if $\tau \in \mathcal{O}^{\star} \setminus \mathcal{O}$ then $-\tau \in \mathcal{O}$, since $-\tau = -1 \cdot \tau \in \mathcal{O}^{\star}$ and $\det(-\tau) = 1$. Thus $\mathcal{O}^{\star} = \mathcal{O} \cup \mathcal{O}(-1)$ and in fact $\mathcal{O}^{\star} = \mathcal{O} \times \{\pm 1\}$.

Proposition 2.5.5 *If $G \leq O(\mathbb{R}^3)$ has rotation subgroup H then either $H = G$, or H has index 2 in G; in particular $H \triangleleft G$.*

Proof Fix $\tau \in G \setminus H$ and take any $\sigma \in G \setminus H$. Then $\det(\sigma \tau^{-1}) = (-1)^2 = 1$, so $\sigma \tau^{-1} \in H$, $\sigma \in H\tau$, and $G = H \cup H\tau$. $\triangle$

If $G \leq O(\mathbb{R}^3)$ with rotation subgroup $H \neq G$ let us distinguish between two cases, viz. $-1 \in G$ and $-1 \notin G$.

If $-1 \in G$ then just as for $\mathcal{O}^{\star}$ above we have $G = H \times \{\pm 1\}$. On the other hand, if H is *any* group of rotations in $O(\mathbb{R}^3)$, we may form $G = H^{\star} = H \times \{\pm 1\}$ and obtain a subgroup G of $O(\mathbb{R}^3)$ having H as its rotation subgroup.

Suppose then that $-1 \notin G$, and say that $H\tau$ is the coset different from H in G. Then it is easy to check (do so!) that the set $K = H \cup H(-\tau)$ is a group of rotations in $O(\mathbb{R}^3)$ having H as a subgroup of index 2.

To turn the tables, suppose that K is a group of rotations in $O(\mathbb{R}^3)$ and that K has a subgroup H of index 2. Set $G = H \cup \{-\tau : \tau \in K \setminus H\}$ and check that G is a subgroup of $O(\mathbb{R}^3)$ having H as its rotation subgroup. We will denote this group by $G = K]H$; as an example $\mathcal{O}]\mathcal{T}$ is the full symmetry group of a regular tetrahedron (why?).

Combining Theorem 2.5.4 with the discussion above we have proved a theorem due originally to the crystallographer J. F. C. Hessel.

Theorem 2.5.6 (Hessel, 1830) *If G is a finite subgroup of $O(\mathbb{R}^3)$ then G is one of*

1. $C_n^{(3)}$, $n \geq 1$; $D_n^{(3)}$, $n \geq 2$; $\mathcal{T}$; $\mathcal{O}$; $\mathcal{I}$;

2. $C_n^{(3)^{\star}}$, $n \geq 1$; $D_n^{(3)^{\star}}$, $n \geq 2$; $\mathcal{T}^{\star}$; $\mathcal{O}^{\star}$; $\mathcal{I}^{\star}$; or

3. $C_{2n}^{(3)}]C_n^{(3)}$, $n \geq 1$; $D_n^{(3)}]C_n^{(3)}$, $n \geq 2$; $D_{2n}^{(3)}]D_n^{(3)}$, $n \geq 2$; $\mathcal{O}]\mathcal{T}$.

Note that there are three groups of order 2 in the theorem, viz. $C_2^{(3)}$, $(C_1^{(3)})^{\star}$, and $C_2^{(3)}]C_1^{(3)}$. As abstract groups they are all, of course, isomorphic. But they are geometrically different — the technical meaning of that is that they are not conjugate as subgroups of $O(\mathbb{R}^3)$. Of course the entire spirit of Hessel's Theorem is to list the finite subgroups of $O(\mathbb{R}^3)$ up to conjugacy. For the subgroups of order 2 the nonidentity elements are, respectively, a rotation through angle π, an inversion through the origin, and a reflection. From another viewpoint they have -1 as an eigenvalue with differing multiplicities 2, 3, and 1.

It is a mildly tedious exercise (not assigned) to check that all the groups listed in Hessel's Theorem are geometrically distinct.

Although it is not the usual definition, it is true that a group G in Hessel's Theorem is a *point group* for a crystal if and only if the orders of elements in G are 1, 2, 3, 4, or 6.

Exercise

There are 32 crystallographic point groups in $O(\mathbb{R}^3)$. Find them.

A final remark: the results of this section are in a sense more general than they may appear to be. If V is a finite dimensional real vector space and G is *any* finite group of invertible linear transformations on V then G is conjugate to a subgroup of the orthogonal group $O(V)$. For a proof see e.g. page 97 of [32]; or, for a complex analogue (whose proof is similar), see Proposition 10.3.13 below.

2.6 Some Linear Groups

If V is an n-dimensional vector space over a field F then $\mathrm{GL}(V)$ denotes the *general linear group* of V, i.e. the group of all invertible linear transformations on V. Choosing a basis for V provides an isomorphism of $\mathrm{GL}(V)$ with the group $\mathrm{GL}(n, F)$ of all invertible $n \times n$ matrices over F.

The determinant is a homomorphism from $\mathrm{GL}(V)$ (or $\mathrm{GL}(n, F)$) onto the multiplicative group $F^\star$; its kernel is the subgroup $\mathrm{SL}(V)$ (or $\mathrm{SL}(n, F)$), the *special linear group*.

If F is finite, with q elements, then the matrix groups are often denoted by $\mathrm{GL}(n, q)$ and $\mathrm{SL}(n, q)$.

Exercises

1. Show that

$$Z(\mathrm{GL}(V)) = \{a1 : a \in F^\star\} \quad \text{and} \quad Z(\mathrm{SL}(V)) = \{a1 : a \in F^\star, a^n = 1\}.$$

2. Let G be the (additive) elementary abelian group $\mathbb{Z}_p^n$ of order p^n, where p is a prime. Show that the holomorph of G is isomorphic with the subgroup of $\mathrm{GL}(n+1, p)$ consisting of all matrices having the partitioned form

$$\begin{bmatrix} A & v \\ 0 & 1 \end{bmatrix},$$

with $A \in \mathrm{GL}(n, p)$ and $v \in G$.

If $0 \neq v \in V$ write $[v]$ for the line Fv through the origin spanned by v, and call it a *projective point*. The set of all distinct projective points $[v]$ is called the *projective space* of dimension $n - 1$ based on V, and is denoted by $P_{n-1}(V)$, or simply by $P(V)$. There is a natural permutation action of $\mathrm{GL}(V)$ on $P(V)$, given by $\tau[v] = [\tau v]$ for all $\tau \in \mathrm{GL}(V)$, $[v] \in P(V)$. It is easy to check that the kernel of the action is $Z(\mathrm{GL}(V))$, and likewise that the kernel of the action of $\mathrm{SL}(V)$ on $P(V)$ is $Z(\mathrm{SL}(V))$.

Exercise

If $\dim V = n$ and $|F| = q$, finite, show that $|P(V)| = (q^n - 1)/(q - 1)$.

Define the *projective general linear group* of V to be

$$\mathrm{PGL}(V) = \mathrm{GL}(V)/Z(\mathrm{GL}(V))$$

and the *projective special linear group* to be

$$\mathrm{PSL}(V) = \mathrm{SL}(V)/Z(\mathrm{SL}(V));$$

each of them acts faithfully on $P(V)$. There are isomorphic copies based on the corresponding matrix groups, with obvious corresponding notation.

Proposition 2.6.1 *If $|F| = q$, finite, then*

1. $|\mathrm{GL}(n, q)| = q^{n(n-1)/2} \prod\{q^k - 1 : 1 \le k \le n\}$,

2. $|\mathrm{SL}(n, q)| = |\mathrm{PGL}(n, q)| = |\mathrm{GL}(n, q)|/(q - 1)$, *and*

3. $|\mathrm{PSL}(n, q)| = |\mathrm{SL}(n, q)|/(n, q - 1)$.

Proof The first row of a matrix in $\mathrm{GL}(n, q)$ can be any of the $q^n - 1$ nonzero row vectors over F; the second must be independent of the first, so there are $q^n - q$ choices, etc. Thus

$$
\begin{aligned}
|\mathrm{GL}(n, q)| &= \prod_1^n (q^n - q^k) = q^{1 + \cdots + (n-1)} \prod_1^n (q^k - 1) \\
&= q^{n(n-1)/2} \prod\{q^k - 1 : 1 \le k \le n\}.
\end{aligned}
$$

Part 2 follows, since $|F^\star| = q - 1$, $Z(\mathrm{GL}(n, q)) \cong F^\star$, and $\mathrm{GL}(n, q)/\mathrm{SL}(n, q) \cong F^\star$. Since $F^\star$ is cyclic it follows from the exercise on page 39 that $a \in Z(\mathrm{SL}(n, q))$ if and only if $a^n = a^{q-1} = 1$, which is if and only if $a^{(n, q-1)} = 1$. In other words, a must be in the unique subgroup of order $(n, q - 1)$ in $F^\star$. $\triangle$

Although SL and PGL have the same order they are usually *not* isomorphic.

Exercise

If F is any field denote by $\mathcal{M}_2(F)$ the group of all *linear fractional transformations* (sometimes called *Möbius transformations*) of F, i.e. all rational functions $f(x) \in F(x)$ of the form

$$f(x) = \frac{ax + b}{cx + d},$$

with $ad - bc \neq 0$. Show that $\mathcal{M}_2(F) \cong \mathrm{PGL}(2, F)$.

Let $\mathcal{L}_2(F)$ be the subgroup of $\mathcal{M}_2(F)$ consisting of those $f(x)$ for which $ad - bc$ is a square in $F^\star$. Show that $\mathcal{L}_2(F) \cong \mathrm{PSL}(2, F)$.

A *hyperplane* in V is any subspace of codimension 1. If $1 \neq \tau \in \mathrm{GL}(V)$ then τ is called a *transvection* if there is a hyperplane W such that $\tau|_W = 1_W$ and $\tau v - v \in W$ for all $v \in V$; W is called the *fixed hyperplane* of τ.

If τ is a transvection with fixed hyperplane W choose a basis for V consisting of first some $v_1 \in V \setminus W$ and then a basis $\{v_2, \ldots, v_n\}$ for W. It is clear from the matrix representing τ relative to this basis that $\det \tau = 1$, so in fact $\tau \in \mathrm{SL}(V)$.

Exercises

1. Show that the inverse of a transvection is a transvection.

2. Suppose that V is a subspace of a space V_1, that $v \in V_1 \setminus V$, and that τ is a transvection on V with fixed hyperplane W. Show that τ can be extended to a transvection τ_1 on V_1 whose fixed hyperplane W_1 contains v.

Proposition 2.6.2 *If u and v are linearly independent in V then there is a transvection τ with $\tau u = v$.*

Proof Choose a hyperplane W in V with $u - v \in W$ but $u \notin W$, and define τ by means of $\tau|_W = 1_W$, $\tau u = v$. If $x \in V$ write $x = au + w$, where $a \in F$ and $w \in W$. Then $\tau x - x = av + w - au - w = a(v - u) \in W$, so τ is a transvection. $\triangle$

Proposition 2.6.3 *Suppose that W_1 and W_2 are two distinct hyperplanes in V and that $v \in V \setminus (W_1 \cup W_2)$. Then there is a transvection τ with $\tau W_1 = W_2$ and $\tau v = v$.*

Proof Note that $W_1 + W_2 = V$, so $\dim W_1 \cap W_2 = n - 2$ and $W = W_1 \cap W_2 + Fv$ is another hyperplane. Write $v = x + y$, with $x \in W_1$ and $y \in W_2$. Then $x \notin W_2$, so $W_1 = W_1 \cap W_2 + Fx$, and likewise $W_2 = W_1 \cap W_2 + Fy$. Thus $V = W_1 \cap W_2 + Fx + Fy$. It follows that $x \notin W$, or else $y = v - x$ is also in W and hence $V \subseteq W$!! Define τ via $\tau|_W = 1_W$ and $\tau x = y$. Then τ is a transvection just as in the proof above, $\tau v = v$ since $v \in W$, and $\tau W_1 = \tau(W_1 \cap W_2 + Fx) = W_1 \cap W_2 + Fy = W_2$. $\triangle$

Theorem 2.6.4 *The set of transvections generates* $\mathrm{SL}(V)$.

Proof Fix $\rho \in \mathrm{SL}(V)$, then choose a hyperplane W in V and choose $v \in V \setminus W$. If v and ρv are linearly independent then by Proposition 2.6.2 there is a transvection τ_1 with $\tau_1 \rho v = v$. If v and ρv are linearly dependent then first choose a transvection τ_0 so that v and $\tau_0 \rho v$ are linearly independent, then a transvection τ_1' so that $\tau_1' \tau_0 \rho v = v$, and set $\tau_1 = \tau_1' \tau_0$. Thus in either case we have $\tau_1 \rho v = v$ and τ_1 is a product of transvections. Note that $v \notin \tau_1 \rho W$. If $\tau_1 \rho W = W$ set $\tau_2 = 1_V$. If $\tau_1 \rho W \neq W$ apply Proposition 2.6.3 to get a transvection τ_2 with $\tau_2 \tau_1 \rho W = W$ and $\tau_2 v = v$. Set $\sigma = \tau_2 \tau_1 \rho$. Since $\sigma v = v$ it follows that $\sigma|_W \in \mathrm{SL}(W)$. Now use induction on $n = \dim V$. If $n = 2$ then $\sigma|_W = 1_W$, so $\sigma = 1$ and $\rho = \tau_1^{-1} \tau_2^{-1}$. If $n > 2$ then by induction $\sigma|_W$ is a product of transvections on W, each of which extends to a transvection fixing v on V by Exercise 2 on page 41. Thus σ is the product of the extended transvections, and $\rho = \tau_1^{-1} \tau_2^{-1} \sigma$, a product of transvections. $\triangle$

Proposition 2.6.5 *If τ_1 and τ_2 are transvections on V then they are conjugate in* $\mathrm{GL}(V)$. *If $n > 2$ they are conjugate in* $\mathrm{SL}(V)$.

Proof For $i = 1, 2$ write W_i for the fixed hyperplane of τ_i, choose $x_i \in V \setminus W_i$, and set $w_i = \tau_i x_i - x_i \in W_i$. Choose bases $\{w_1, u_3, \ldots u_n\}$ for W_1 and $\{w_2, v_3, \ldots, v_n\}$ for W_2. For each $a \in F^\star$ define $\sigma_a \in \mathrm{GL}(V)$ by setting $\sigma_a x_1 = x_2$, $\sigma_a w_1 = w_2$, $\sigma_a u_i = v_i$ for $3 \leq i \leq n - 1$, and, if $n > 2$, $\sigma_a u_n = a v_n$. Then a straightforward calculation verifies that $\sigma_a \tau_1 \sigma_a^{-1}$ and τ_2 agree on $x_2, w_2, u_3, \ldots, u_n$, and hence $\sigma_a \tau_1 \sigma_a^{-1} = \tau_2$.

If $n > 2$ we may set $b = (\det \sigma_1)^{-1}$ and obtain $\sigma_b \in \mathrm{SL}(V)$. $\triangle$

Proposition 2.6.6 *Suppose that* $\dim V = 2$, *and let $\{v_1, v_2\}$ be any basis for V. Every transvection is conjugate in $\mathrm{SL}(V)$ to one whose matrix relative to $\{v_1, v_2\}$ is of the form* $\begin{bmatrix} 1 & 0 \\ a & 1 \end{bmatrix}$, $a \in F^\star$.

Proof If τ is a transvection with fixed hyperplane (line) W choose $v \in V \setminus W$ and set $w = \tau v - v \in W$. Relative to the basis $\{v, w\}$ the matrix representing τ is $\begin{bmatrix} 1 & 0 \\ 1 & 1 \end{bmatrix}$. If M is the matrix that represents τ relative to $\{v_1, v_2\}$ then there is a matrix B in $\mathrm{GL}(2, F)$ with $B^{-1} M B = \begin{bmatrix} 1 & 0 \\ 1 & 1 \end{bmatrix}$. If $\det B = a^{-1}$ we may set $A = \begin{bmatrix} a & 0 \\ 0 & 1 \end{bmatrix}$. Then $BA \in \mathrm{SL}(2, F)$, and

$$(BA)^{-1} M (BA) = A^{-1} \begin{bmatrix} 1 & 0 \\ 1 & 1 \end{bmatrix} A = \begin{bmatrix} 1 & 0 \\ a & 1 \end{bmatrix}.$$

$\triangle$

Theorem 2.6.7 *If $n \geq 3$ and $G = \mathrm{SL}(V)$ then $G' = G$.*

Proof By Theorem 2.6.4 and Proposition 2.6.5 it will suffice to exhibit a transvection in G'. Choose a basis $\{v_1, \ldots, v_n\}$ for V and define τ_1, τ_2 via

$$\tau_1 : v_1 \mapsto v_1 - v_2, \; v_i \mapsto v_i \; \text{ if } 2 \leq i \leq n,$$

$$\tau_2 : v_1 \mapsto v_1, \; v_2 \mapsto v_2 - v_3, \; v_i \mapsto v_i \; \text{ if } 3 \leq i \leq n.$$

Then

$$\tau_1 \tau_2 \tau_1^{-1} \tau_2^{-1} : v_1 \mapsto v_1 - v_3, \; v_i \mapsto v_i \; \text{ if } 2 \leq i \leq n,$$

so $\tau_1 \tau_2 \tau_1^{-1} \tau_2^{-1}$ is a transvection in G'. △

Corollary 2.6.8 *If $n \geq 3$ then $\mathrm{PSL}(V)$ is equal to its derived group.*

Theorem 2.6.9 *If $n = 2$, $|F| > 3$, and $G = \mathrm{SL}(V)$, then $G' = G$.*

Proof Choose a basis $\{v_1, v_2\}$ for V and choose $a \in F^\star$, $a \neq \pm 1$. Define $\sigma \in \mathrm{SL}(V)$ via $\sigma(v_1) = a^{-1}v_1$, $\sigma(v_2) = av_2$, and for each $b \in F^\star$ define $\tau_b \in \mathrm{SL}(V)$ via $\tau_b(v_1) = v_1 + bv_2$, $\tau_b(v_2) = v_2$. Then $\sigma \tau_b \sigma^{-1} \tau_b^{-1}$ is represented by the matrix

$$\begin{bmatrix} a^{-1} & 0 \\ 0 & a \end{bmatrix} \begin{bmatrix} 1 & 0 \\ b & 1 \end{bmatrix} \begin{bmatrix} a & 0 \\ 0 & a^{-1} \end{bmatrix} \begin{bmatrix} 1 & 0 \\ -b & 1 \end{bmatrix} = \begin{bmatrix} 1 & 0 \\ ba(a - a^{-1}) & 1 \end{bmatrix}.$$

The theorem follows from Theorem 2.6.4 and Proposition 2.6.6, since $b \in F^\star$ is arbitrary. △

Corollary 2.6.10 *If $n = 2$ and $|F| > 3$ then $\mathrm{PSL}(V)$ is equal to its derived group.*

Exercise

If $n \geq 3$ or $|F| > 3$ show that the derived group of $\mathrm{GL}(V)$ is $\mathrm{SL}(V)$. Determine the derived groups of $\mathrm{SL}(2,2)$, $\mathrm{SL}(2,3)$, and $\mathrm{GL}(2,3)$.

Proposition 2.6.11 *If $n \geq 2$ then $\mathrm{PSL}(V)$ acts faithfully and doubly transitively on the projective space $P(V)$. In particular it is primitive.*

Proof It was observed earlier that $\mathrm{PSL}(V)$ is faithful on $P(V)$. Take $[v_1] \neq [v_2]$ and $[w_1] \neq [w_2]$ in $P(V)$, so $\{v_1, v_2\}$ and $\{w_1, w_2\}$ are linearly independent sets in V. If $n = 2$ they are bases. If $n \geq 3$ set $V_1 = Fv_1 + Fv_2$ and $V_2 = Fw_1 + Fw_2$. Then either $V_1 = V_2 \neq V$ or else $V_1 \neq V_2$, in which case $V_1 \cup V_2$ is not a subspace – in either case $V_1 \cup V_2 \neq V$. If $v_3 \in V \setminus (V_1 \cup V_2)$ then both $\{v_1, v_2, v_3\}$ and $\{w_1, w_2, v_3\}$ are linearly independent. The argument may be repeated to obtain two bases for V, $\{v_1, v_2, v_3, \ldots v_n\}$ and $\{w_1, w_2, v_3, \ldots, v_n\}$.

For any $b \in F^\star$ define $\tau_b \in \mathrm{GL}(V)$ via $v_1 \mapsto bw_1$, $v_2 \mapsto w_2$, $v_i \mapsto v_i$ for $3 \le i \le n$. If $w_j = \sum_i a_{ij} v_i$, $j = 1, 2$, then $\det \tau_b = b(a_{11}a_{22} - a_{12}a_{21})$. Choose b so that $\det \tau_b = 1$; then $\tau_b \in \mathrm{SL}(V)$, and it carries $[v_1]$ to $[w_1]$, $[v_2]$ to $[w_2]$. $\triangle$

Proposition 2.6.12 *If $0 \ne v \in V$ and $A = \mathrm{Stab}_{\mathrm{SL}(V)}([v])$ then A has an abelian normal subgroup B whose conjugates in $\mathrm{SL}(V)$ generate $\mathrm{SL}(V)$.*

Proof Choose a hyperplane W with $v \notin W$, so $V = W + Fv$. If $\sigma \in A$ and $w \in W$ write $\sigma w = \sigma' w + a_w v$, with $\sigma' w \in W$ and $a_w \in F$. It is easy to check that $\sigma' \in \mathrm{GL}(W)$ and that $\varphi \colon \sigma \mapsto \sigma'$ is a homomorphism from A into $\mathrm{GL}(W)$. Thus $B = \ker \varphi$ is a normal subgroup of A.

Choose a basis $\{w_1, \ldots, w_{n-1}\}$ for W. For any $b \in F^\star$ define $\tau_b \in A$ via $w_1 \mapsto w_1 + bv$, $w_i \mapsto w_i$ for $2 \le i \le n - 1$, and $v \mapsto v$. Then τ_b is a transvection (with hyperplane spanned by $\{w_2, \ldots, w_{n-1}, v\}$), and if $w \in W$ then $\tau_b w = w + bv$, so $\tau_b' = 1_W$ and $\tau_b \in B$. If $n = 2$ the matrix representing τ_b relative to the basis $\{w_1, v\}$ is $\begin{bmatrix} 1 & 0 \\ b & 1 \end{bmatrix}$, and $b \in F^\star$ is arbitrary. It follows from Theorem 2.6.4 and Propositions 2.6.5 and 2.6.6 that the conjugates of B generate $\mathrm{SL}(V)$ for all $n \ge 2$. If $\sigma \in B$ then $\sigma' = 1_W$, so $\sigma w_i = w_i + a_i v$, all i, $a_i \in F$. Thus, if $\sigma_1, \sigma_2 \in B$, their representing matrices relative to the basis $\{w_1, \ldots, w_{n-1}, v\}$ have the partitioned form

$$\begin{bmatrix} I & 0 \\ u_1 & 1 \end{bmatrix} \quad \text{and} \quad \begin{bmatrix} I & 0 \\ u_2 & 1 \end{bmatrix}.$$

Since

$$\begin{bmatrix} I & 0 \\ u_1 & 1 \end{bmatrix} \begin{bmatrix} I & 0 \\ u_2 & 1 \end{bmatrix} = \begin{bmatrix} I & 0 \\ u_1 + u_2 & 1 \end{bmatrix} = \begin{bmatrix} I & 0 \\ u_2 & 1 \end{bmatrix} \begin{bmatrix} I & 0 \\ u_1 & 1 \end{bmatrix}$$

we see that B is abelian. $\triangle$

Theorem 2.6.13 *Except for $\mathrm{PSL}(2,2)$ and $\mathrm{PSL}(2,3)$, every $\mathrm{PSL}(V)$ is a simple group.*

Proof Choose $v \ne 0$ in V, and take $A = \mathrm{Stab}_{\mathrm{SL}(V)}([v])$ and $B \triangleleft A$ as in Proposition 2.6.12. Write $Z = Z(\mathrm{SL}(V))$ and set $H = A/Z = \mathrm{Stab}_{\mathrm{PSL}(V)}([v])$, $K = BZ/Z \triangleleft H$. Then all the conditions of Iwasawa's Theorem (1.6.9) are met, and $\mathrm{PSL}(V)$ is simple. $\triangle$

Exercise

Use the action on the projective line P to show that $\mathrm{PSL}(2,2) \cong \mathrm{Sym}(3)$, $\mathrm{PSL}(2,3) \cong \mathrm{Alt}(4)$, and $\mathrm{PSL}(2,4) \cong \mathrm{Alt}(5)$. Note also that $\mathrm{PSL}(2,5) \cong \mathrm{Alt}(5)$ (why?).

Note that $|PSL(3,4)| = 20160 = |Alt(8)|$, and that both groups are simple. Let us show that they are *not* isomorphic. For any $\sigma \in SL(3,4)$ write $\bar{\sigma}$ for its image in PSL.

If $\bar{\sigma}$ is any element of order 2 in $PSL(3,4)$ then $\sigma^2 \in Z(SL(3,4))$, so $\sigma^6 = 1$. Replacing σ by $\tau = \sigma^3$ we have $\bar{\tau} = \bar{\sigma}$ and τ of order 2 in SL as well as in PSL. Let $W = \ker(1+\tau) \subseteq V$, so $\dim V = 3 = \dim W + \dim(1+\tau)V$. Note that $v \in W$ if and only if $\tau v = v$ (characteristic 2), so $W \neq V$. Also $\tau(1+\tau)v = (\tau+1)v$, all $v \in V$; i.e. $(1+\tau)v$ is fixed by τ. Thus $(1+\tau)V \subseteq W$. Now $\dim(1+\tau)V \leq \dim W$, and it follows that $\dim(1+\tau)V = 1$, $\dim W = 2$. Consequently W is a hyperplane and $\tau v - v \in W$ for all $v \in V$; i.e. τ is a transvection. By Proposition 2.6.5 we see that all elements of order 2 are conjugate in $PSL(3,4)$.

Since $Alt(8)$ has nonconjugate elements of order 2, e.g. $(1\,2)(3\,4)$ and $(1\,2)(3\,4)(5\,6)(7\,8)$, it is immediate that $PSL(3,4) \not\cong Alt(8)$.

Exercise

Note also that $PSL(4,2)$ and $PSL(3,4)$ are of the same order. Show that $PSL(4,2) \not\cong PSL(3,4)$ by considering the matrices

$$\begin{bmatrix} 1 & 0 & 0 & 0 \\ 0 & 1 & 0 & 0 \\ 0 & 0 & 1 & 0 \\ 0 & 0 & 1 & 1 \end{bmatrix} \quad \text{and} \quad \begin{bmatrix} 1 & 0 & 0 & 0 \\ 1 & 1 & 0 & 0 \\ 0 & 0 & 1 & 0 \\ 0 & 0 & 1 & 1 \end{bmatrix}$$

in $SL(4,2)$.

It is in fact true that $PSL(4,2) \cong Alt(8)$. One proof proceeds by finding a set of generators for $PSL(4,2)$ that satisfy the relations in a presentation for $Alt(8)$; see Huppert [37], page 157, for details.

It can be shown that any simple group of order 168 is isomorphic to $PSL(2,7)$ (see [37], page 184). That applies in particular to $PSL(3,2)$, the collineation group of the world's smallest projective plane P, with 7 points and 7 lines.

Finally, $|PSL(2,9)| = 360 = |Alt(6)|$. In this case $PSL(2,9)$ has a subgroup isomorphic with $PSL(2,5)$, hence of index 6, so the action on its cosets gives an isomorphism into $Sym(6)$, and $PSL(2,9) \cong Alt(6)$ since it is the unique subgroup of index 2 of $Sym(6)$. Again, see [37] for details.

The argument used in the proof of Proposition 2.6.11 shows that $PGL(V)$ acts faithfully and doubly transitively on the projective space $P(V)$. Note that $|PGL(2,q)| = (q+1)q(q-1)$, a product of three consecutive integers, and that the projective line $P = P_2(q)$ has $q+1$ points. A glance at page 17 suggests the possibility of sharp 3-transitivity.

Theorem 2.6.14 *If* $\dim V = 2$ *then* $PGL(V)$ *is sharply triply transitive on the projective line* $P = P(V)$.

Proof Write G for $\mathrm{PGL}(V)$, and if $\tau \in \mathrm{GL}(V)$ write $\bar{\tau}$ for its image in G. We know that G is doubly transitive on P. For 3-transitivity it will suffice to choose $[u] \neq [v]$ in P and to show that the 2-point stabilizer $G_{[u],[v]}$ is transitive on $P' = P \setminus \{[u], [v]\}$. Since $\{u, v\}$ is a basis for V every point in P' can be written as $[au + bv]$, where $a, b \in F$ and $ab \neq 0$. Take any such point $[w] \in P'$, $w = au + bv$. Define $\tau \in \mathrm{GL}(V)$ by setting $\tau u = au$ and $\tau v = bv$. Then $\bar{\tau} \in G$ and $\bar{\tau}([u + v]) = [au + bv] = [w]$, and the G-orbit of $[u + v]$ is all of P'.

For sharpness take $w = u + v$, so $[w] \neq [u]$ or $[v]$, and suppose that $\bar{\tau} \in G_{[u],[v],[w]}$. Say that $\tau u = au$, $\tau v = bv$, and $\tau w = cw$. Then

$$cu + cv = cw = \tau(u + v) = \tau u + \tau v = au + bv,$$

so $a = c = b$ and $\bar{\tau} = 1 \in G$. $\triangle$

2.7   Mathieu Groups

We begin this section with some ideas about permutation groups that could well have gone into Chapter 1.

Suppose that G acts transitively on a set S and $\alpha \notin S$. Set $S^{\star} = S \cup \{\alpha\}$ and extend the action of G from S to $S^{\star}$ by simply agreeing that $\alpha^x = \alpha$, all $x \in G$. The hope is to embed G within a larger group $G^{\star}$ acting transitively on $S^{\star}$ and having $G = \mathrm{Stab}_{G^{\star}}(\alpha)$. If that can be done then $G^{\star}$ is called a *one-point extension* of G. Note that if G is (sharply) k-transitive on S, and such a $G^{\star}$ exists, then $G^{\star}$ is (sharply) $(k + 1)$-transitive on $S^{\star}$.

For a rather obvious example take $G = S_n$, $S = \{1, 2, \ldots, n\}$, $\alpha = n + 1$, and $G^{\star} = S_{n+1}$. Unfortunately, interesting one-point extensions are rather rare. The next theorem, while somewhat technical, provides the possibility of some interesting constructions.

Theorem 2.7.1 (E. Witt, 1938) *Suppose that G is faithful and k-transitive on S, $k \geq 2$, and $S^{\star} = S \cup \{\alpha\}$, with $\alpha \notin S$. Set $\alpha^x = \alpha$, all $x \in G$. Suppose that there are elements $x \in G$, $h \in \mathrm{Perm}(S^{\star})$, and $s, t \in S$, satisfying*

1. $s^x = t$ and $t^x = s$;

2. $s^h = \alpha$, $\alpha^h = s$, and $t^h = t$;

3. $(xh)^3 \in G$ and $h^2 \in G$; and

4. $hG_s h = G_s$.

Then $G^{\star} = \langle G, h \rangle$ is a one-point extension of G.

Proof Note that $x \notin G_s$, so $G = G_s \cup G_s x G_s$ by Corollary 1.6.3. Let us show that $hGh \subseteq G \cup GhG$. Since $s^{h^2} = \alpha^h = s$ and $h^2 \in G$ we have

$h^2 \in G_s$, and by part 4 $hG_s = G_s h$. Also $(xh)^3 = (xhx)(hxh) \in G$, so $hxh \in x^{-1}h^{-1}G = x^{-1}h^{-1}h^2 G = x^{-1}hG$. Thus

$$
\begin{aligned}
hGh &= h(G_s \cup G_s x G_s)h = hG_s h \cup hG_s x G_s h \\
&= G_s \cup G_s hxhG_s \subseteq G \cup G_s x^{-1} hGG_s \subseteq G \cup GhG.
\end{aligned}
$$

It follows that $G \cup GhG$ is a group, for $GhG \cdot GhG = G \cdot hGh \cdot G \subseteq G \cup GhG$ and $(GhG)^{-1} = Gh^{-1}G = Gh^{-1}h^2 G = GhG$, and in fact $G \cup GhG = \langle G, h \rangle = G^\star$. But then $G_\alpha^\star$ is clearly G, since no element of GhG stabilizes α. $\quad\triangle$

If F is a field and $V = F^2$ then the points of the projective line $P = P(V)$ are of two types: first there is the set of points $[(a, 1)]$, $a \in F$, which can be identified with F itself via $[(a, 1)] \leftrightarrow a$, and there is one additional point $[(1, 0)]$, commonly called the *point at infinity* and denoted ∞. With that identification $P = F \cup \{\infty\}$.

Recall now the group $\mathcal{L}_2(F)$ of linear fractional transformations in the exercise on page 41. It is isomorphic with $\mathrm{PSL}(2, F)$, which acts doubly transitively on the projective line. If $f(x) = (ax + b)/(cx + d) \in \mathcal{L}_2(F)$ then $f(x)$ determines a function from $F \cup \{\infty\}$ to itself via substitution, with $f(\infty) = a/b$ ($= \infty$ if $b = 0$), etc. As a result $\mathcal{L}_2$ acts on $P = F \cup \{\infty\}$, and it is easy to check that the isomorphism with $\mathrm{PSL}(2, F)$ provides an equivalence of actions. In particular, $\mathcal{L}_2$ is faithful and doubly transitive on P by Proposition 2.6.11.

If F is finite and of odd characteristic then $F^\star$ is cyclic of even order, and the set $F^{\times 2}$ of squares in $F^\star$ is just half of $F^\star$ (each $a^2 \in F^{\times 2}$ has two square roots, viz. $\pm a$).

Suppose now that $F = \mathbb{F}_9$. Recall that the map $x \mapsto x^3$ is an automorphism of $\mathbb{F}_9$. We enlarge $\mathcal{L}_2(9)$ by adjoining the transformations $g(x) = (ax^3 + b)/(cx^3 + d)$, where $a, b, c, d \in F$ and $ad - bc \in F^\star \setminus F^{\times 2}$, and call the resulting set M_{10}.

Exercise

Show that M_{10} is a group having $\mathcal{L}_2(9)$ as a subgroup of index 2.

The action of $\mathcal{L}_2(9)$ on the 10-point projective line $P = \mathbb{F}_9 \cup \{\infty\}$ extends naturally to an action of M_{10} on P (also by substitution), it is at least doubly transitive since $\mathcal{L}_2$ is doubly transitive. The 2-point stabilizer $(M_{10})_{0,\infty}$ consists of the transformations of the form $f(x) = ax^\epsilon$, $a \in F^\star$, with $\epsilon = 1$ if $a \in F^{\times 2}$, $\epsilon = 3$ otherwise. Clearly the $(M_{10})_{0,\infty}$-orbit of 1 is all of $F^\star$, so M_{10} is 3-transitive on P. If $f(x) = ax^\epsilon$ is in the 3-point stabilizer $(M_{10})_{0,1,\infty}$ then $1 = f(1) = a$, and $\epsilon = 1$ since $1 \in F^{\times 2}$; i.e. $f(x)$ is the identity element of M_{10}, and M_{10} is sharply triply transitive. Its order is $10 \cdot 9 \cdot 8 = 720$.

The field $\mathbb{F}_9$ is a degree 2 extension of $\mathbb{F}_3$, so it is generated over $\mathbb{F}_3$ by a root of an irreducible quadratic. Since $x^2 + x - 1$ is irreducible over $\mathbb{F}_3$ we may take $\mathbb{F}_9 = \mathbb{F}_3(\delta)$, with $\delta^2 = 1 - \delta$. Thus

$$
\mathbb{F}_9 = \{a + b\delta : a, b \in \mathbb{F}_3\}.
$$

Exercise

Define g_1 and g_2 in M_{10} via $g_1(x) = \delta^2 x$ and $g_2(x) = \delta x^3$. Show that $\langle g_1, g_2 \rangle$ is a quaternion group of order 8.

We are now in a position to carry out the procedure of Witt's Theorem (2.7.1). Choose some point $\alpha \notin P = \mathbb{F}_9 \cup \{\infty\}$ and set $P^\star = P \cup \{\alpha\}$. Define $h \in \text{Perm}(P^\star)$ by setting

$$h(\alpha) = \infty, \ h(\infty) = \alpha, \ \text{and} \ h(a + b\delta) = a - b\delta$$

for $a, b \in \mathbb{F}_3$. Take $f \in M_{10}$ defined by $f(x) = 1/x$; it will play the role of x in Witt's Theorem. The roles of s and t are played, then, by ∞ and 0, and the first two conditions of the theorem are met. Clearly $h^2 = 1$, and fh has orbits $\{1\}$, $\{-1\}$, $\{0, \infty, \alpha\}$, $\{\delta, -1 - \delta, 1 - \delta\}$, and $\{-\delta, 1 + \delta, -1 + \delta\}$, so also $(fh)^3 = 1$.

The stabilizer $(M_{10})_\infty$ has order 72; it contains the translation subgroup T consisting of all f_a, $a \in \mathbb{F}_9$, where $f_a(x) = x + a$, and the two transformations g_1 and g_2, with $g_1(x) = \delta^2 x$ and $g_2(x) = \delta x^3$. It follows from the exercise above that $(M_{10})_\infty = \langle g_1, g_2, f_a : a \in \mathbb{F}_9 \rangle$, so to verify condition 4 in Witt's Theorem it will suffice to conjugate those generators by h. Straightforward calculations show that $hg_1h = g_2g_1$, $hg_2h = g_2^{-1}$, and if $a = b + c\delta \in \mathbb{F}_9$ then $hf_ah = f_{b - c\delta}$.

Set $M_{11} = \langle M_{10}, h \rangle$, the *first Mathieu group*.

We may repeat the procedure and extend M_{11}. Choose a point $\beta \notin P^\star$ and set $P^{\star\star} = P^\star \cup \{\beta\} = \mathbb{F}_9 \cup \{\infty, \alpha, \beta\}$. Define $k \in \text{Perm}(P^{\star\star})$ via

$$k(\alpha) = \beta, \ k(\beta) = \alpha, \ k(\infty) = \infty, \quad \text{and} \quad k(x) = x^3 \ \text{for} \ x \in \mathbb{F}_9.$$

It is clear that the first two conditions of Witt's Theorem are met, with x, h, s, t, and α replaced, respectively, by h, k, α, ∞, and β. Also $k^2 = 1$, and the orbits of hk are

$$\{0\}, \ \{1\}, \ \{-1\}, \ \{\infty, \alpha, \beta\}, \ \{\delta, -1 + \delta, 1 + \delta\}, \ \{-\delta, 1 - \delta, -1 - \delta\},$$

so $(hk)^3 = 1$. The stabilizer $(M_{11})_\alpha$ is M_{10}. For $f \in M_{10}$, say with $f(x) = (ax^\epsilon + b)/(cx^\epsilon + d)$, we have $kfk(x) = (a^3 x^\epsilon + b^3)/(c^3 x^\epsilon + d^3)$, and $a^3 d^3 - b^3 c^3 = (ad - bc)^3 \in F^{\times 2}$ if and only if $ad - bc \in F^{\times 2}$, so $kfk \in M_{10}$.

Set $M_{12} = \langle M_{11}, k \rangle$, the *second Mathieu group*. We have proved

Theorem 2.7.2 *The first Mathieu group M_{11} is a one-point extension of M_{10}; it is sharply 4-transitive of degree 11, and has order $11 \cdot 10 \cdot 9 \cdot 8 = 7,920 = 2^4 \cdot 3^2 \cdot 5 \cdot 11$. The second Mathieu group M_{12} is a one-point extension of M_{11}; it is sharply 5-transitive of degree 12, and has order $12 \cdot 11 \cdot 10 \cdot 9 \cdot 8 = 95,040 = 2^6 \cdot 3^3 \cdot 5 \cdot 11$.*

To begin another round of one-point extensions take $F = \mathbb{F}_4$ and $G = \text{PSL}(3,4)$. If δ is a primitive element for F then $\delta^3 = 1$, $\delta^2 = 1 + \delta$, and $F = \mathbb{F}_2(\delta) = \{0, 1, \delta, 1 + \delta\}$.

As representative vectors in $V = F^3$ for the points of the projective plane $P = P_2(V)$ we may take

$$v_{a,b} = (a, b, 1), \quad v_a = (a, 1, 0), \quad \text{and} \quad v_\infty = (1, 0, 0),$$

with $a, b \in F$, so $|P| = 21$. Recall that G is doubly transitive on P (Proposition 2.6.11). Choose a point $\alpha \notin P$ and set $P^\star = P \cup \{\alpha\}$, as usual.

Since $a \mapsto a^2$ is an automorphism of F the map f_1 defined by $f_1(a, b, c) = (a^2 + bc, b^2, c^2)$ is a permutation of F^3, and it induces a permutation g_1 of P via $g_1[v] = [f_1 v]$. Set $s_1 = [v_\infty]$ and $t_1 = [v_0]$ in P. Define $h_1 \in \text{Perm}(P^\star)$ via $h(\alpha) = s_1$, $h(s_1) = \alpha$, and $h_1 = g_1$ on $P \setminus \{s_1\}$; note that $h_1(t_1) = t_1$ and $h_1^2 = 1$.

Next define $x_1 \in G$ via $x_1: [(a, b, c)] \mapsto [(b, a, c)]$. Check that

$$(x_1 h_1)^3: [(a, b, c)] \mapsto [(ac + b^2(1 + c^3), bc + a^2(1 + c^3), c^2)].$$

If $c \neq 0$ the image is $[(ac, bc, c^2)] = [(a, b, c)]$; if $c = 0$ but $ab \neq 0$ we may take $b = 1$ and $a = 1, \delta, \delta^2$, and check that each point is fixed by $(x_1 h_1)^3$. Similarly s_1, t_1, and α are fixed, and we conclude that $(x_1 h_1)^3 = 1$.

If $y \in G_{s_1}$ then $y = \overline{Y}$, with

$$Y = \begin{bmatrix} 1 & u & x \\ 0 & v & y \\ 0 & w & z \end{bmatrix} \in \text{SL}(3, 4),$$

so $\det Y = vz - yw = 1$. Thus $f_1 Y f_1$ maps (a, b, c) to

$$(a + b(u^2 + vw) + c(x^2 + yz), bv^2 + cy^2, bw^2 + cz^2),$$

which is the result of multiplication by the matrix

$$Z = \begin{bmatrix} 1 & u^2 + vw & x^2 + yz \\ 0 & v^2 & y^2 \\ 0 & w^2 & z^2 \end{bmatrix} \in \text{SL}(3, 4).$$

Also $h_1 y h_1$ fixes α, so it follows that $h_1 G_{s_1} h_1 = G_{s_1}$.

The conditions of Witt's Theorem are thus met, and we define the *third Mathieu group* M_{22} to be the one-point extension $\langle G, h_1 \rangle$. It has order $22 \cdot |\text{PSL}(3,4)| = 443,520$.

Keeping the notation above, choose a point $\beta \notin P^\star$ and set $P^{\star\star} = P^\star \cup \{\beta\}$. Define f_2 on F^3 via $f_2(a, b, c) = (a^2, b^2, c^2\delta)$, then $g_2 \in \text{Perm}(P)$ via $g_2[v] = [f_2 v]$. Set $s_2 = \alpha$ and $t_2 = [v_\infty]$ in $P^\star$, and define $h_2 \in \text{Perm}(P^{\star\star})$ by setting $h_2(\beta) = s_2$, $h_2(s_2) = \beta$, $h_2 = g_2$ on P; note that $h_2(t_2) = t_2$ and

$h_2^2 = 1$. Set $x_2 = h_1 \in M_{22}$. It is straightforward (using $f_1 f_2$) to calculate that $(x_2 h_2)^3 = 1$.

Note that $(M_{22})_{s_2} = (M_{22})_\alpha = G = \mathrm{PSL}(3,4)$, and that $G = G_{s_1} \cup G_{s_1} x_1 G_{s_1}$ by Corollary 1.6.3. Thus to show that $h_2 (M_{22})_{s_2} h_2 = (M_{22})_{s_2}$ it will suffice to show that $h_2 G_{s_1} h_2 \le G$ and that $h_2 x_1 h_2 \in G$. For the first of these we may use $y = \overline{Y} \in G_{s_1}$, as above, and calculate that the effect of $f_2 Y f_2$ on F^3 is the same as multiplication by

$$\begin{bmatrix} 1 & u^2 & x^2 \delta^2 \\ 0 & v^2 & y^2 \delta^2 \\ 0 & w^2 \delta & z^2 \end{bmatrix} \in \mathrm{SL}(3,4),$$

so $h_2 y h_2 \in G$. Finally $h_2 x_1 h_2 = x_1$.

Define next the *fourth Mathieu group* M_{23} to be the one-point extension $\langle M_{22}, h_2 \rangle$ of order $23 \cdot |M_{22}| = 10,200,960$.

One more time. Choose $\gamma \notin P^{\star\star}$ and set $P^{\star\star\star} = P^{\star\star} \cup \{\gamma\}$. Define f_3 on F^3 by setting $f_3(a,b,c) = (a^2, b^2, c^2)$, then g_3 on P via $g_3[v] = [f_3 v]$. Define $h_3 \in \mathrm{Perm}(P^{\star\star\star})$ via $h_3(\gamma) = \beta$, $h_3(\beta) = \gamma$, $h_3(\alpha) = \alpha$, and $h_3 = g_3$ on P. Set $x_3 = h_2 \in M_{23}$, $s_3 = \beta$, $t_3 = \alpha$, and observe that $(M_{23})_{s_3} = M_{22} = \langle G, h_1 \rangle = \langle G_{s_1}, x_1, h_1 \rangle$. It is easy now to check that $h_3^2 = (x_3 h_3)^3 = 1$, $h_3 G_{s_1} h_3 = G_{s_1}$, $h_3 x_1 h_3 = x_1$, and $h_3 h_1 h_3 = h_1$. Thus we may define the *fifth Mathieu group* M_{24} to be the one-point extension $\langle M_{23}, h_3 \rangle$, of order $24 \cdot |M_{23}| = 244,823,040$.

We summarize in a theorem.

Theorem 2.7.3 *The third Mathieu group M_{22} is a one-point extension of $\mathrm{PSL}(3,4)$; it is 3-transitive of degree 22 and has order $443,520 = 2^7 \cdot 3^2 \cdot 5 \cdot 7 \cdot 11$. The fourth Mathieu group M_{23} is a one-point extension of M_{22}; it is 4-transitive of degree 23 and has order $10,200,960 = 2^7 \cdot 3^2 \cdot 5 \cdot 7 \cdot 11 \cdot 23$. The fifth Mathieu group M_{24} is a one-point extension of M_{23}; it is 5-transitive of degree 24 and has order $244,823,040 = 2^{10} \cdot 3^3 \cdot 5 \cdot 7 \cdot 11 \cdot 23$.*

The groups M_{12} and M_{24} do not admit further transitive extensions. A proof can be found in [38].

Since M_{11} has degree 11 (and is faithful) we may view it as a subgroup of $\mathrm{Sym}(11)$, acting on $S = \{1, 2, \dots, 11\}$. Similar remarks apply to the other Mathieu groups as well. In fact, a bit later we will exhibit generators for them in the appropriate symmetric groups.

Theorem 2.7.4 *The five Mathieu groups are all simple.*

Proof We begin with $G = M_{11}$. Choose $E \in \mathrm{Syl}_{11}(G)$; it is cyclic of order 11, so a generator for it must be an 11-cycle and E is transitive. Set $C = C_G(E) \ge E$. Take $\tau \in \mathrm{Stab}_C(1)$. For any $i \in S = \{1, \dots, 11\}$ choose $\sigma \in E$ with $1^\sigma = i$ and observe that $i^\tau = 1^{\sigma\tau} = 1^{\tau\sigma} = 1^\sigma = i$, so $\tau = 1$. Thus C is regular on S, so $|C| = 11$ and $C = E$. Next set $N = N_G(E)$; N acts by conjugation on E as a group of automorphisms. The kernel of the action is $C = E$, and $|\operatorname{Aut}(E)| = 10$, so if $k = [N:E]$ then $k \mid 10$ and $|N| = 11k$. The number of distinct Sylow 11-subgroups of G is $\ell = [G:N] \equiv 1 \pmod{11}$, so $|G| = 11k\ell = 11 \cdot 10 \cdot 9 \cdot 8$. This forces $k = 5$ and $\ell = 144$.

Suppose now that $1 \neq K \lhd G$. Then K is transitive on S by Proposition 1.6.7, so $11 \mid |K|$ and we may assume $E \leq K$. In fact, since $K \lhd G$, K must contain *all* of the 144 Sylow 11-subgroups of G, and $|K|$ is a multiple of $11 \cdot 144 = 11 \cdot 2 \cdot 9 \cdot 8$; thus $[G:K]$ is 1 or 5. If $K \neq G$ then $5 \nmid |K|$, so $N \cap K = N_K(E) = E$. By a theorem of Burnside, which we borrow from a later chapter (Theorem 4.1.9), there is a characteristic subgroup H of K (hence $H \lhd G$) such that $K = H \rtimes E$. But now $11 \nmid |H|$ is in conflict with $H \lhd G$ because of Proposition 1.6.7, so in fact $K = G$ and $G = M_{11}$ is simple.

The others are now easy. Since the stabilizer of a point in M_{12} is M_{11}, which is simple, then if M_{12} were not simple it would have a regular normal subgroup by Theorem 1.6.10. But that contradicts Theorem 1.6.12, since M_{12} is 5-transitive. The stabilizer of a point in M_{22} is $PSL(3,4)$, also simple. Thus if M_{22} were not simple it would have a regular normal subgroup, by Theorem 1.6.10, and would have degree 2^k or 3 by Theorem 1.6.12, whereas the degree is 22. The same reasoning applies to M_{23} and M_{24}. $\triangle$

There is much more to be said about the Mathieu groups. They were discovered in 1861 by E. Mathieu [46], although his arguments were not totally convincing. In fact C. Jordan and G. A. Miller tried to argue in subsequent papers that M_{24} could not exist. In 1938 E. Witt (see [66]) gave essentially the constructions that appear above, and he also discussed the groups as automorphism groups of combinatorial objects called *Steiner triple systems* (see [67], also [4]). The groups M_{12} and M_{24} are also automorphism groups of certain *Golay codes* (see [2]).

For over a hundred years the Mathieu groups were the only known simple groups that did not fit into infinite families such as the alternating groups and projective special linear groups; as such Burnside called them *sporadic* groups. Further sporadic simple groups were discovered beginning in the 1960s – the list is now complete, there are all told 26 of them. It is a consequence of the classification of finite simple groups (see [30]) that M_{12} and M_{24} are the only nontrivial examples of 5-fold transitive groups.

For purposes of calculation it is useful to have (permutation) generators for the groups. The ones given below are a slight variation on generators given by Mathieu himself in [47]. Set

$$\lambda = (3, 7, 11, 8)(4, 10, 5, 6),$$

$$\mu = (1, 2, 3, 4, 5, 6, 7, 8, 9, 10, 11),$$

$$\nu = (1, 12)(2, 11)(3, 6)(4, 8)(5, 9)(7, 10),$$

$$\pi = (1, 21, 10, 7, 8, 12, 9)(2, 19, 15, 6, 11, 18, 20)(3, 16, 17, 13, 4, 5, 14),$$

$$\rho = (1, 6, 20, 12, 10, 21, 11)(2, 8, 13, 15, 5, 16, 22)(3, 4, 17, 19, 9, 14, 18),$$

$$\sigma = (1, 2, 3, 4, 5, 6, 7, 8, 9, 10, 11, 12, 13, 14, 15, 16, 17, 18, 19, 20, 21, 22, 23),$$

$$\tau_1 = (1, 24)(2, 23)(3, 12)(4, 16)(5, 18)(6, 10),$$

$$\tau_2 = (7, 20)(8, 14)(9, 21)(11, 17)(13, 22)(15, 19),$$

and $\tau = \tau_1 \tau_2$. Then $M_{11} = \langle \lambda, \mu \rangle$, $M_{12} = \langle \lambda, \mu, \nu \rangle$, $M_{22} = \langle \pi, \rho \rangle$, $M_{23} = \langle \pi, \sigma \rangle$, and $M_{24} = \langle \pi, \tau \rangle$.

Exercise

If F is a finite field show that $\text{Aff}(F)$, acting on F, has $\mathcal{M}_2(F)$ as a one-point extension, the new point being ∞ in $P = F \cup \{\infty\}$. Use this information to reprove Theorem 2.6.14, at least for finite fields.

2.8   Symplectic Groups

Suppose that V is a vector space of dimension $n = 2m$ over a field F, and let B be a nondegenerate alternate form on V. An ordered basis $\{u_1, v_1, \ldots, u_m, v_m\}$ for V built up of hyperbolic pairs $\{u_i, v_i\}$, as in Theorem 1.7.10, is called a *symplectic* basis for V, V is called a *symplectic space*, and somewhat loosely speaking the study of (V, B) is called *symplectic geometry*.

An invertible linear transformation τ of V is said to be *symplectic* if $B(\tau v, \tau w) = B(v, w)$ for all $v, w \in V$, *i.e.* if τ is an isometry of V. The set of all symplectic transformations of V is evidently a subgroup of $\text{GL}(V)$; it is called the *symplectic group* on V and denoted $\text{Sp}(V)$. The dependence of $\text{Sp}(V)$ on the form B is not significant. Since any other nondegenerate alternate form B_1 on V is equivalent with B (Corollary 1.7.12), the resulting symplectic group (relative to B_1) is conjugate to $\text{Sp}(V)$ in $\text{GL}(V)$. In other words, relative to appropriately chosen bases the two groups are represented by the *same* matrices in $\text{GL}(n, F)$.

Suppose, then, that a symplectic basis has been chosen for V. If $\tau \in \text{GL}(V)$ say that τ is represented, relative to that basis, by the matrix T. An easy calculation (as on page 20) shows that $\tau \in \text{Sp}(V)$ if and only if $T^t \widehat{B} T = \widehat{B}$. The resulting group of matrices is denoted $\text{Sp}(n, F)$, or also by $\text{Sp}(n, q)$ if $|F| = q$, finite.

Proposition 2.8.1 *If* $\dim V = 2$ *then* $\text{Sp}(V) = \text{SL}(V)$.

Proof We may calculate with $\text{Sp}(2, F)$ via a symplectic basis, hence

$$\widehat{B} = M = \begin{bmatrix} 0 & 1 \\ -1 & 0 \end{bmatrix}.$$

By the remarks above $T = \begin{bmatrix} a & b \\ c & d \end{bmatrix} \in \mathrm{GL}(2, F)$ is in $\mathrm{Sp}(n, F)$ if and only if $T^t M T = M$, i.e.

$$\begin{bmatrix} 0 & ad - bc \\ -ad + bc & 0 \end{bmatrix} = \begin{bmatrix} 0 & 1 \\ -1 & 0 \end{bmatrix}.$$

$\triangle$

Exercise

Let $\{u_1, v_1, \ldots, u_m, v_m\}$ be a symplectic basis for V, but re-order the vectors as

$$\{u_1, u_2, \ldots, u_m, v_1, v_2, \ldots, v_m\}.$$

Determine the resulting representing matrix for B. Show that every

$$\begin{bmatrix} A & 0 \\ 0 & (A^t)^{-1} \end{bmatrix}, \quad A \in \mathrm{GL}(m, F),$$

represents a symplectic transformation relative to this basis, and conclude that $\mathrm{Sp}(2m, F)$ has a subgroup isomorphic to $\mathrm{GL}(m, F)$.

Recall that if $1 \neq \tau \in \mathrm{SL}(V)$, then τ is called a *transvection*, with fixed hyperplane W, if $\tau|_W = 1_W$ and $\tau v - v \in W$ for all $v \in V$.

Let τ be a transvection with fixed hyperplane W, and suppose that $\tau \in \mathrm{Sp}(V)$. By Proposition 1.7.4 dim $W^\perp = 1$, so say that $W^\perp = \mathrm{Span}(u)$. By Corollary 1.7.5 we have $u^\perp = W$, but also $u \in u^\perp = (W^\perp)^\perp = W$. Choose $x \in V \setminus W$, so $V = \mathrm{Span}(x) \oplus W$. Then $0 \neq \tau x - x = z$, say, in W. For any $v \in V$ write $v = bx + w$, $b \in F$ and $w \in W$, and define $f \in V^*$ via $f(v) = b$; thus ker $f = W$. By Corollary 1.7.2 $\exists y \in V$ such that $f(v) = B(v, y)$, all $v \in V$. Note that $\tau v = \tau(bx + w) = b\tau x + w = b(x + z) + w = v + bz = v + f(v)z = v + B(v, y)z$, all $v \in V$. In particular if $w \in W$ then $w = \tau w = w + B(w, y)z$, so $B(w, y) = 0$, all w, and $y \in W^\perp = \mathrm{Span}(u)$. Say that $y = cu$, $c \in F^*$. Also, since $\tau \in Sp(V)$, $B(w, x) = B(\tau w, \tau x) = B(w, x + z) = B(w, x) + B(w, z)$, so $B(w, z) = 0$ for all $w \in W$, and $z \in W^\perp$, say $z = du$, $d \in F^*$. Thus $\tau v = v + B(v, cu) \cdot du = v + cdB(v, u)u$. Set $a = cd$ and we have

$$\tau v = v + aB(v, u)u,$$

all $v \in V$, i.e. τ is determined by the scalar a and the vector u. Write $\tau = \tau_{u,a}$, and call τ a *symplectic transvection* with direction u.

Conversely, given any nonzero $u \in V$ and any $a \in F^*$ we may define $\tau = \tau_{u,a}$ via $\tau v = v + aB(v, u)u$, all v. Then clearly τ is a transvection with fixed hyperplane $W = u^\perp$, and it is easy to check directly that $B(\tau v, \tau w) = B(v, w)$, all $v, w \in V$, so $\tau_{u,a} \in \mathrm{Sp}(V)$.

It will be convenient to extend the notation slightly and define $\tau_{u,a}$ by the formula above even when $a = 0$. Of course $\tau_{u,0}$ is simply the identity transformation.

Exercise

Fix $u \neq 0$ in V.

Show that $\tau_{u,a}\tau_{u,b} = \tau_{u,a+b}$, $\tau_{bu,a} = \tau_{u,ab^2}$, and $\tau_{u,a}^{-1} = \tau_{u,-a}$.

2. Conclude that $\{\tau_{u,a}: a \in F\}$ is a group isomorphic with the additive group F.

3. If $\sigma \in \mathrm{Sp}(V)$ then $^{\sigma}\tau_{u,a} = \tau_{\sigma u,a}$.

If V is a symplectic space write $\mathcal{T} = \langle \tau_{u,a}: 0 \neq u \in V,\ a \in F^{\star}\rangle$, the subgroup of $\mathrm{Sp}(V)$ generated by symplectic transvections. It will be shown eventually that $\mathcal{T} = \mathrm{Sp}(V)$, so $\mathrm{Sp}(V)$ is generated by symplectic transvections.

Proposition 2.8.2 *If V is symplectic then the group $\mathcal{T} \leq \mathrm{Sp}(V)$ is transitive on $V \setminus \{0\}$.*

Proof Take $v \neq w$ in $V\setminus\{0\}$. If $B(v,w) \neq 0$ set $a = 1/B(v,w)$ and $u = v-w$. Then

$$\tau_{u,a}(v) = v + aB(v,u)u = v + \frac{B(v,v-w)}{B(v,w)}(v-w) = v - (v-w) = w.$$

Suppose then that $B(v,w) = 0$. Choose $f \in V^*$ with $f(v) \neq 0$ and $f(w) \neq 0$. By Corollary 1.7.2 $\exists u \in V$ such that $f(v) = B(v,u) \neq 0$ and $f(w) = B(w,u) \neq 0$. By the first part of the proof $\exists \tau_1, \tau_2 \in \mathcal{T}$ with $\tau_1 v = u$ and $\tau_2 u = w$, hence $\tau_2\tau_1 v = w$. △

Recall that a *hyperbolic pair* in V is an ordered pair (u,v) of vectors with $B(u,v) = 1$.

Proposition 2.8.3 *If V is symplectic then the group $\mathcal{T} \leq \mathrm{Sp}(V)$ is transitive on the set $\mathcal{S}$ of all hyperbolic pairs in V.*

Proof Take $(u_1,v_1), (u_2,v_2) \in \mathcal{S}$. First use Proposition 2.8.2 to choose $\tau \in \mathcal{T}$ with $\tau u_1 = u_2$, and hence $\tau: (u_1,v_1) \mapsto (u_2,v_3) \in \mathcal{S}$, $v_3 = \tau v_1$. Now we need $\sigma \in \mathcal{T}$ with $\sigma: (u_2,v_3) \mapsto (u_2,v_2)$, i.e. $\sigma \in \mathrm{Stab}_{\mathcal{T}}(u_2)$ and $\sigma v_3 = v_2$. If $B(v_3,v_2) \neq 0$ use $\sigma = \tau_{u,a}$, with $u = v_3 - v_2$, as in the proof of Proposition 2.8.2. Then $\sigma v_3 = v_2$ and also $B(u_2,v_3) = 1 = B(u_2,v_2)$, so $B(u_2,u) = B(u_2,v_3-v_2) = 0$, and therefore $\sigma u_2 = u_2 + aB(u_2,u)u = u_2$.

Suppose then that $B(v_3,v_2) = 0$. Note that $(u_2,u_2+v_3) \in \mathcal{S}$, and that $B(v_3,u_2+v_3) \neq 0$, so by what was just proved $\exists \sigma_1 \in \mathcal{T}$ with $\sigma_1 u_2 = u_2$ and $\sigma_1 v_3 = u_2 + v_3$. But also $B(u_2+v_3,v_2) = B(u_2,v_2) \neq 0$, so $\exists \sigma_2 \in \mathcal{T}$ with $\sigma_2 u_2 = u_2$ and $\sigma_2(u_2+v_3) = v_2$. Setting $\sigma = \sigma_2\sigma_1 \in \mathcal{T}$ we have $\sigma u_2 = u_2$ and $\sigma v_3 = \sigma_2(u_2+v_3) = v_2$. Finally, $\sigma\tau \in \mathcal{T}$ and $\sigma\tau: (u_1,v_1) \mapsto (u_2,v_2)$. △

Theorem 2.8.4 *The symplectic group $\mathrm{Sp}(V)$ is generated by its symplectic transvections.*

Proof Use induction on m, where $2m = n = \dim V$. The case $m = 1$ is covered by Proposition 2.8.1 and Theorem 2.6.4. Choose a hyperbolic pair (u, v) in V and set $W = \mathrm{Span}(u, v)$, a hyperbolic plane. Then $V = W \oplus W^{\perp}$ by Proposition 1.7.9. Take any $\sigma \in \mathrm{Sp}(V)$. Then $(\sigma u, \sigma v)$ is a hyperbolic pair, so by Proposition 2.8.3 $\exists \tau \in \mathcal{T}$ with $\tau \sigma u = u$ and $\tau \sigma v = v$, so $\tau \sigma|_W = 1_W$. But also $\tau \sigma|_{W^{\perp}} \in \mathrm{Sp}(W^{\perp})$. By induction $\tau \sigma|_{W^{\perp}}$ is a product of symplectic transvections on $W^{\perp}$. Any such transvection, say $\tau_{w,a}$, determines, by the *same* formula, viz. $\tau_{w,a}(v) = v + aB(v, w)w$, a symplectic transvection on V whose fixed hyperplane contains W. Since also $\tau \sigma|_W = 1_W$ we conclude that $\tau \sigma \in \mathcal{T}$, and hence that $\sigma \in \mathcal{T}$. $\triangle$

Corollary 2.8.5 $\mathrm{Sp}(V) \leq \mathrm{SL}(V)$.

Proposition 2.8.6 *The center of* $\mathrm{Sp}(V)$ *is* $\{\pm 1\}$.

Proof Note that $\tau_{u,a} v = v$ if and only if $B(v, u) = 0$, so $u^{\perp}$ is the space of fixed points for $\tau_{u,a}$. If $\sigma \in Z(\mathrm{Sp}(V))$ then in particular σ commutes with every $\tau_{u,a}$. If $v \in u^{\perp}$ then $\tau_{u,a}(\sigma v) = \sigma \tau_{u,a} v = \sigma v$, so $\sigma v \in u^{\perp}$, i.e. $\sigma u^{\perp} = u^{\perp}$ for all u. But $\mathrm{Span}(u) = \mathrm{rad}\, u^{\perp}$ and σ is symplectic, so σ carries $\mathrm{Span}(u)$ to $\mathrm{Span}(u)$; in particular, $\sigma u = c_u u$ for some $c_u \in F$, all $u \in V$. Choose u and v linearly independent in V. Then $\sigma(u + v) = c_{u+v}(u + v) = c_u u + c_v v$, so $c_u = c_{u+v} = c_v = c$, say, independent of u and v, and $\sigma = c1_V$. But then $B(u, v) = B(\sigma u, \sigma v) = c^2 B(u, v)$ for all u, v, so $c = \pm 1$. $\triangle$

Proposition 2.8.7 *If* $|F| \geq 4$ *then the derived group of* $\mathrm{Sp}(V)$ *is* $\mathrm{Sp}(V)$.

Proof Fix $u \neq 0$ in V and $a \in F^{\star}$. Choose $b \in F \setminus \{0, \pm 1\}$ and set $c = a/(1 - b^2)$, $d = -b^2 c$. Then $c + d = a$, so $\tau_{u,c} \tau_{u,d} = \tau_{u,a}$ by the exercise on page 54. Choose $\sigma \in \mathrm{Sp}(V)$ with $\sigma u = bu$ (Proposition 2.8.2). Then $\sigma \tau_{u,c}^{-1} \sigma^{-1} = \sigma \tau_{u,-c} \sigma^{-1} = \tau_{\sigma u,-c} = \tau_{bu,-c} = \tau_{u,-b^2 c} = \tau_{u,d}$, and hence $\tau_{u,c} \sigma \tau_{u,c}^{-1} \sigma^{-1} = \tau_{u,c} \tau_{u,d} = \tau_{u,a} \in \mathrm{Sp}(V)'$. Apply Theorem 2.8.4. $\triangle$

Proposition 2.8.8 *If* $|F| = 3$ *and* $\dim V \geq 4$ *then the derived group of* $\mathrm{Sp}(V)$ *is* $\mathrm{Sp}(V)$.

Proof Choose a symplectic basis $\{u_1, v_1, \ldots, u_m, v_m\}$ and define linear transformations σ and τ as follows:

$$\sigma: u_1 \mapsto u_1 + u_2,\ v_1 \mapsto v_2,\ u_2 \mapsto u_1,\ v_2 \mapsto v_1 - v_2,\ \text{ and}$$

$$u_i \mapsto u_i,\ v_i \mapsto v_i\ \text{ if } i > 2;$$

$$\tau: u_1 \mapsto u_1 - v_1 + v_2,\ v_1 \mapsto v_1,\ u_2 \mapsto u_2 + v_1,\ v_2 \mapsto v_2,\ \text{ and}$$

$$u_i \mapsto u_i,\ v_i \mapsto v_i\ \text{ if } i > 2.$$

It is an exercise to show that σ and τ are in $\mathrm{Sp}(V)$ and to show that $\sigma \tau \sigma^{-1} \tau^{-1} = \tau_{v_1,1}$. Thus $\tau_{v_1,1} \in \mathrm{Sp}(V)'$. Given any $u \neq 0$ in V, choose

$\gamma \in \mathrm{Sp}(V)$ with $\gamma v_1 = u$; then $\gamma \tau_{v_1,1} \gamma^{-1} = \tau_{\gamma v_1,1} = \tau_{u,1} \in \mathrm{Sp}(V)'$ and also $\tau_{u,1}^{-1} = \tau_{u,-1} \in \mathrm{Sp}(V)'$. Apply Theorem 2.8.4. $\triangle$

The case $|F| = 3$ and $\dim V = 2$ is an exception, since $\mathrm{Sp}(2,3) = \mathrm{SL}(2,3)$ of order 24.

Proposition 2.8.9 *If $|F| = 2$ and $\dim V \geq 6$ then the derived group of $\mathrm{Sp}(V)$ is $\mathrm{Sp}(V)$.*

Proof The proof is exactly as for Proposition 2.8.8, except that

$$\sigma: u_1 \mapsto u_1 + u_3, \; v_1 \mapsto v_3, \; u_2 \mapsto u_1, \; v_2 \mapsto v_1 + v_3,$$

$$u_3 \mapsto u_2, \; v_3 \mapsto v_2, \quad \text{and}$$

$$u_i \mapsto u_i, \; v_i \mapsto v_i \; \text{ if } i > 3;$$

$$\tau: u_1 \mapsto u_1 + u_2, \; v_1 \mapsto v_1, \; u_2 \mapsto v_1 + u_2 + v_2 + v_3, \; v_2 \mapsto v_2,$$

$$u_3 \mapsto v_2 + u_3 + v_3, \; v_3 \mapsto v_3, \quad \text{and}$$

$$u_i \mapsto u_i, \; v_i \mapsto v_i \; \text{ if } i > 3.$$

$\triangle$

Since $\mathrm{Sp}(2,2) = \mathrm{SL}(2,2) \cong S_3$ it is an exception to Proposition 2.8.9. It will be shown later that $\mathrm{Sp}(4,2) \cong S_6$, so it is also an exception.

Since $\mathrm{Sp}(V) \leq \mathrm{GL}(V)$ there is an action of $\mathrm{Sp}(V)$ on the projective space $P = P_{n-1}(V)$. The kernel of the action is $Z(\mathrm{Sp}(V)) = \{\pm 1\}$, and we define the *projective symplectic group* $\mathrm{PSp}(V)$ to be $\mathrm{Sp}(V)/Z(\mathrm{Sp}(V))$; it acts faithfully and transitively (see Proposition 2.8.2) on P.

Proposition 2.8.10 *$\mathrm{Sp}(V)$ is primitive on $P = P_{n-1}(V)$.*

Proof We may assume that $n \geq 4$ by Proposition 2.8.1, since $\mathrm{SL}(2,F)$ is doubly transitive. Suppose that $S \subseteq P$, $|S| > 1$, and either $\sigma S = S$ or $\sigma S \cap S = \varnothing$ for each $\sigma \in \mathrm{Sp}(V)$. We show first that there are $[u]$, $[v] \in S$ with $B(u,v) \neq 0$. Suppose to the contrary that $B(u,v) = 0$ for all $[u]$, $[v] \in S$. Choose $[u] \neq [v]$ in S and choose $f \in V^*$ with $f(u) = 1$, $f(v) = 0$. By Corollary 1.7.2 $\exists x \in V$ with $B(u,x) = 1$, $B(v,x) = 0$. Set $W = \mathrm{Span}(u,x)$, a hyperbolic plane, and set $H = \{\sigma \in \mathrm{Sp}(V): \sigma|_W = 1_W\} \leq \mathrm{Sp}(V)$. Since $V = W \oplus W^\perp$ it is clear that every $\sigma \in \mathrm{Sp}(W^\perp)$ extends trivially to $\sigma' \in \mathrm{Sp}(V)$ with $\sigma'|_W = 1_W$, so in fact $\mathrm{Sp}(W^\perp) = \{\tau|_{W^\perp}: \tau \in H\}$. Choose $w \neq 0$ in $W^\perp$. Since $v \in W^\perp$ there is some $\tau \in H$ with $\tau v = w$, by Proposition 2.8.2. Since $\tau u = u$ we have $[u] \in S \cap \tau S$, so $\tau S = S$. Also $[w] = \tau[v] \in S$, and w was an arbitrary nonzero vector in $W^\perp$. Since $W^\perp \neq 0$ there is a hyperbolic pair (y,z) in $W^\perp$; hence $[y]$, $[z] \in S$, but $B(y,z) = 1$, a contradiction.

Thus we may in fact choose $[u]$, $[v] \in S$ with $B(u,v) \neq 0$, so we may assume that $B(u,v) = 1$ and (u,v) is a hyperbolic pair. Now take any $[w] \in P$. If $B(u,w) \neq 0$ we may assume that (u,w) is a hyperbolic pair and by

Proposition 2.8.3 find $\sigma \in \mathrm{Sp}(V)$ with $\sigma u = u$ and $\sigma v = w$. Thus $u \in S \cap \sigma S$, so $\sigma S = S$ and $[w] \in S$. On the other hand, if $B(u, w) = 0$ choose $f \in V^*$ with $f(u) = f(w) = 1$ and thereby obtain $x \in V$ with $B(u, x) = B(w, x) = 1$. As above, $[x] \in S$, and also $\exists \tau \in \mathrm{Sp}(V)$ with $\tau u = w$ and $\tau x = x$. Again $\tau S = S$, since $[x] \in S \cap \tau S$, and $[w] = \tau[u] \in S$. Thus $S = P$.   △

Theorem 2.8.11 *Except for* $\mathrm{PSp}(2, 2)$, $\mathrm{PSp}(2, 3)$, *and* $\mathrm{PSp}(4, 2)$ *every projective symplectic group* $\mathrm{PSp}(V)$ *is simple.*

Proof We know that $\mathrm{PSp}(V)$ acts faithfully on $P = P_{n-1}(V)$, and the action is primitive by Proposition 2.8.10. With the exceptions noted, $\mathrm{PSp}(V)$ is its own derived group by Propositions 2.8.7, 2.8.8, and 2.8.9. Fix $[u] \in P$ and set $H = \mathrm{Stab}_{\mathrm{Sp}(V)}([u])$, $\overline{H} = H/\{\pm 1\} = \mathrm{Stab}_{\mathrm{PSp}(V)}([u])$. Set $K = \{\tau_{u,a} : a \in F\}$. By the exercise on page 54 $K \triangleleft H$ and $K \cong F$, so K is abelian. If $\sigma \in \mathrm{Sp}(V)$ then $^\sigma K \supseteq \{\tau_{\sigma u, a} : a \in F\}$, so $\cup\{^\sigma K : \sigma \in \mathrm{Sp}(V)\}$ contains all symplectic convections by Proposition 2.8.2, and hence generates $\mathrm{Sp}(V)$ by Theorem 2.8.4. Thus $\langle \overline{^\sigma K} : \overline{\sigma} \in \mathrm{PSp}(V)\rangle = \mathrm{PSp}(V)$, and with the exceptions noted $\mathrm{PSp}(V)$ is simple by Iwasawa's Theorem (1.6.9).   △

Theorem 2.8.12 *If* $|F| = q$, *finite, then*

$$|\mathrm{Sp}(n, q)| = q^{m^2} \prod_{i=1}^{m}(q^{2i} - 1),$$

where $n = 2m$, *and*

$$|\mathrm{PSp}(n, q)| = \frac{|\mathrm{Sp}(n, q)|}{(2, q - 1)}.$$

Proof Any one of the $q^n - 1$ nonzero vectors can serve as the first vector u in a hyperbolic pair. Since $|q^\perp| = q^{n-1}$ there are $q^n - q^{n-1}$ vectors v with $B(u, v) \neq 0$. Thus there are $(q^n - q^{n-1})/(q - 1) = q^{n-1}$ choices of v for which $B(u, v) = 1$, and consequently $(q^n - 1)q^{n-1}$ distinct hyperbolic pairs. If (u, v) is a hyperbolic pair and W is the hyperbolic plane with $\{u, v\}$ as basis, then $\sigma \in \mathrm{Stab}_{\mathrm{Sp}(V)}((u, v))$ if and only if $\sigma|_W = 1_W$, and we saw in the proof of Proposition 2.8.10 that $\mathrm{Stab}_{\mathrm{Sp}(V)}((u, v)) \cong \mathrm{Sp}(W^\perp)$. Thus $|\mathrm{Sp}(n, q)| = (q^n - 1)q^{n-1}|\mathrm{Sp}(n - 2, q)|$ by Proposition 1.2.1, and we may argue by induction. When $n = 2$ then $|\mathrm{Sp}(2, q)| = |\mathrm{SL}(2, q)| = q(q^2 - 1)$. Assume inductively that $|\mathrm{Sp}(n - 2, q)| = q^{(m-1)^2} \prod_{i=1}^{m-1}(q^{2i} - 1)$, then

$$|\mathrm{Sp}(n, q)| = q^{2m-1}(q^{2m} - 1)q^{(m-1)^2} \prod_{i=1}^{m-1}(q^{2i} - 1) = q^{m^2} \prod_{1}^{m}(q^{2i} - 1).$$

The relation to $|\mathrm{PSp}(n, q)|$ is immediate from Proposition 2.8.6.   △

Note, for example, that $|\mathrm{Sp}(4, 2)| = 720 = |S_6|$. Let us prepare to show that $\mathrm{Sp}(4, 2) \cong S_6$.

If V has dimension 4 over $F = \mathbb{F}_2$ then $|V| = 16$. Choose $v_1 \neq 0$ in V, and then choose $v_2 \in V \setminus v_1^{\perp}$. Then $v_2 \in v_2^{\perp} \setminus v_1^{\perp}$, so $v_1^{\perp} + v_2^{\perp} = V$, and therefore $\dim(v_1^{\perp} \cap v_2^{\perp}) = 3 + 3 - 4 = 2$ and $|v_1^{\perp} \cap v_2^{\perp}| = 4$. Thus $|v_1^{\perp} \cup v_2^{\perp}| = 8 + 8 - 4 = 12$, and we may choose $v_3 \in V \setminus \{v_1^{\perp} \cup v_2^{\perp}\}$. Similar inclusion-exclusion arguments show that $|v_1^{\perp} \cup v_2^{\perp} \cup v_3^{\perp}| = 14$, so $\exists v_4 \in V \setminus \{v_1^{\perp} \cup v_2^{\perp} \cup v_3^{\perp}\}$. Then $|v_1^{\perp} \cup v_2^{\perp} \cup v_3^{\perp} \cup v_4^{\perp}| = 15$, and the last remaining element v_5 of V is in $V \setminus \{v_1^{\perp} \cup v_2^{\perp} \cup v_3^{\perp} \cup v_4^{\perp}\}$. The resulting set $\{v_1, \ldots, v_5\}$ satisfies $B(v_i, v_j) = 1$ if $i \neq j$, and is a maximal nonorthogonal set in V.

Now we approach maximal nonorthogonal sets from another direction. Choose a hyperbolic pair (v_1, v_2) in V and set $W = \mathrm{Span}(v_1, v_2)$. Then $V = W \oplus W^{\perp}$ and we may choose a hyperbolic pair (x_1, x_2) in $W^{\perp}$. If we set $v_3 = v_1 + v_2 + x_1$, $v_4 = v_1 + v_2 + x_2$, and $v_5 = v_1 + v_2 + x_1 + x_2$, then it is a routine matter to check that $\{v_1, \ldots, v_5\}$ is a maximal nonorthogonal set (m.n.s.).

It is in fact the *unique* m.n.s. containing $\{v_1, v_2\}$. The reason: given (v_1, v_2), suppose that we try to adjoin w so that $\{v_1, v_2, w\}$ is a nonorthogonal set. Write $w = a_1 v_1 + a_2 v_2 + b_1 x_1 + b_2 x_2$. Then $1 = B(v_1, w) = a_2$ and $1 = B(v_2, w) = a_1$, and $w = v_1 + v_2 + b_1 x_1 + b_2 x_2$. If we try taking $b_1 = b_2 = 0$, giving, say, $w' = v_1 + v_2$, then $w' \perp w$ for any other choice of values of b_1 and b_2, so $\{v_1, v_2, w'\}$ is not contained in a larger nonorthogonal set. There remain only the other three possible assignments of values to b_1 and b_2.

By the proof of Theorem 2.8.12 there are $(2^4 - 1) \cdot 2^{4-1} = 120$ hyperbolic pairs in V. Each m.n.s. contains $5 \cdot 4 = 20$ hyperbolic pairs, so there are exactly $120/20 = 6$ distinct m.n.s.'s in V.

Proposition 2.8.13 $\mathrm{Sp}(4, 2) \cong S_6$.

Proof Let S be the set of 6 m.n.s.'s in V, as above. It follows from Proposition 2.8.3 that $\mathrm{Sp}(4, 2)$ acts transitively on S, so there is a homomorphism $\varphi : \mathrm{Sp}(4, 2) \to S_6$. Let $K = \ker \varphi$. Choose a hyperbolic pair (v_1, v_2), set $W = \mathrm{Span}(v_1, v_2)$, and choose a hyperbolic pair (x_1, x_2) in $W^{\perp}$. As above we obtain two m.n.s.'s $\{v_i\}$ and $\{x_i\}$, with

$$v_3 = v_1 + v_2 + x_1, \quad v_4 = v_1 + v_2 + x_2, \quad v_5 = v_1 + v_2 + x_1 + x_2;$$

$$x_3 = x_1 + x_2 + v_1, \quad x_4 = x_1 + x_2 + v_2, \quad x_5 = x_1 + x_2 + v_1 + v_2.$$

Set $z = v_1 + v_2 + x_1 + x_2$ and note that $\{v_i\} \cap \{x_i\} = \{z\}$. Every $\sigma \in K$ fixes $\{v_i\}$ and $\{x_i\}$, so σ fixes $\{v_i\} \cap \{x_i\}$; i.e. $\sigma z = z$, all $\sigma \in K$. But $\mathrm{Sp}(V)$ is transitive on $P(V) = V \setminus \{0\}$, and for each $\tau \in \mathrm{Sp}(V)$ we have ${}^{\tau}\sigma(\tau z) = \tau z$; i.e. τz is fixed by all ${}^{\tau}\sigma \in K$, which is all of K. Since $\mathrm{Sp}(V) = \mathrm{PSp}(V)$ acts faithfully on $P(V)$ we conclude that $K = 1$. Thus φ is 1–1, and hence is an isomorphism since $|\mathrm{Sp}(V)| = |S_6|$. $\triangle$

Thus $\mathrm{Sp}(4, 2)$ is not simple, but its derived group $(\cong A_6)$ is simple and of index 2.

Exercises

1. Suppose that V is a space with an alternate bilinear form B and that W is a subspace. Set $U = W/\operatorname{rad} W$. Show that U is a symplectic space if we define

$$\overline{B}(w_1 + \operatorname{rad} W, w_2 + \operatorname{rad} W) = B(w_1, w_2).$$

2. (This is an alternate approach to Proposition 2.8.13). Take V of dimension 6 over $F = \mathbb{F}_2$ with basis $\{v_1, \ldots, v_6\}$. Define B via $B(v_i, v_j) = 1 + \delta_{ij}$, and verify that (V, B) is a symplectic space. Observe that S_6 is embedded isomorphically in $\operatorname{Sp}(V)$ by permuting $\{v_1, \ldots, v_6\}$. Set $W = \operatorname{Span}(v_1 + \cdots + v_6)$; W and $W^\perp$ are S_6-invariant, so S_6 acts on $U = W^\perp/W$, a 4-dimensional symplectic space by Exercise 1 above. The transposition $(1\,2) \in S_6$ embeds as the transvection $\tau_{v_1+v_2,1}$ in $\operatorname{Sp}(V)$ and maps to a transvection in $\operatorname{Sp}(U)$. Its conjugates in S_6 yield all 15 transvections in $\operatorname{Sp}(U)$, and therefore $S_6 \cong \operatorname{Sp}(U)$. Provide the details.

Note that $|\operatorname{PSp}(4,3)| = 3^4 \cdot (3^2 - 1)(3^4 - 1)/2 = 25,920$. A. Cayley and G. Salmon proved in the 1840s that each sufficiently general cubic surface has 27 straight lines lying on it. C. Jordan (who first defined the symplectic groups) determined the group of incidence-preserving permutations of the 27 lines, and found it to have order 51,840. The group is in fact the Weyl group of type E_6 (e.g. see [32] or [36]), and it has a copy of $\operatorname{PSp}(4,3)$ as a subgroup of index 2.

In classical plane geometry it was known that each sufficiently general quartic curve has 28 "bitangents," i.e. lines that are tangent to the curve at two distinct points. This seems to have first been proved rigorously by K. Jacobi in 1850. Jordan showed that the symmetry group of the 28 bitangents is a simple group of order 1,451,520 – it is isomorphic with $\operatorname{PSp}(6,2)$.

For further discussion of the history of these configurations and relations between them, and for further references, see [63]. Also see the Atlas [14], where the groups are denoted $S_4(3)$ and $S_6(2)$.

Chapter 3

Counting with Groups

This chapter is concerned with enumeration techniques for orbits under group actions. The techniques afford a wide variety of applications to combinatorics, and thereby to many other disciplines. For example, one of the early sources [51] was concerned with enumerating chemical polymers.

Except for the Orbit Formula (Proposition 3.1.2), which is used later, the chapter is by and large independent of the other chapters.

3.1 Fixed Points and Orbits

Suppose the group G acts on a set S (assume that both are finite). If $x \in G$ and $s \in S$ then s is a *fixed point* for x if $s^x = s$. Write $\mathrm{Fix}(x)$ for the set of all fixed points in S of x. Define the *character* θ of the action via $\theta(x) = |\mathrm{Fix}(x)|$, all $x \in G$. Thus θ is a function from G to $\mathbb{N}$.

Proposition 3.1.1 *The character θ of a group action is a class function on G, i.e. θ is constant on conjugacy classes of G.*

Proof If $x, y \in G$ verify that $\mathrm{Fix}(x^y) = \mathrm{Fix}(x)^y$, and so

$$\theta(x^y) = |\mathrm{Fix}(x^y)| = |\mathrm{Fix}(x)^y| = |\mathrm{Fix}(x)| = \theta(x).$$

$\triangle$

Proposition 3.1.2 (The Orbit Formula) *If G acts on S with character θ then the number of distinct G-orbits in S is*

$$k = |G|^{-1} \sum \{\theta(x) : x \in G\}.$$

Proof Set $\mathcal{S} = \{(s, x) \in S \times G : s^x = s\}$. For fixed $s \in S$ there are $|\mathrm{Stab}_G(s)|$ such ordered pairs; for fixed $x \in G$ there are $\theta(x)$ of them. Thus

$$|\mathcal{S}| = \sum \{|\mathrm{Stab}_G(s)| : s \in S\} = \sum \{\theta(x) : x \in G\}.$$

To evaluate the first sum let $\mathcal{O}_1, \ldots, \mathcal{O}_k$ be the distinct G-orbits in S, and choose $s_i \in \mathcal{O}_i$, $1 \le i \le k$. Then

$$\sum \{|\operatorname{Stab}_G(s)| : s \in S\} = \sum_{i=1}^{k} |\mathcal{O}_i| |\operatorname{Stab}_G(s_i)| = k|G|$$

by Proposition 1.2.1, so $k|G| = \sum_x \theta(x)$. $\triangle$

Corollary 3.1.3 *If G is transitive on S, and $|S| > 1$, then G has an element x having no fixed points.*

Proof We have $1 = |G|^{-1} \sum \{\theta(x) : x \in G\}$, which can be interpreted as saying that the average number of fixed points for elements of G is 1, so not all elements can have more than 1 fixed point. In fact, there is an element ($x = 1$) that has more than 1 fixed point, so to compensate there must be at least one element having fewer than 1 (therefore 0) fixed points. $\triangle$

The Orbit Formula is often referred to as *Burnside's Orbit Formula*, e.g. in [31], probably because it appeared in Burnside's book [10]. However it was proved earlier by Frobenius, and a special case of it even earlier by Cauchy. See [49] and [68] for a discussion of the history.

For an easy example of the use of the Orbit Formula imagine coloring the faces of a regular tetrahedron using 2 colors, say red (r) and blue (b), each face being a solid color. Since each of the 4 faces can be one of 2 colors there are $2^4 = 16$ distinct ways the coloring can be done. Distinct does not mean *distinguishable*, however. For example any one of the 4 ways of coloring with 1 r and 3 b's can be made to look exactly like any other one by simply rotating the tetrahedron appropriately. Clearly any 2 colorings are indistinguishable precisely when one can be rotated to the other.

Something slightly subtle has happened. We are no longer talking about the action of the rotation group $\mathcal{T}$ on the set of 4 faces, but on the set of 16 colorings of the faces. To count distinguishable colorings we need to count the orbits under that action. To that end we count first fixed points of group elements. The identity 1 has of course 16 fixed points. There are just 2 other types of rotation in $\mathcal{T}$; there are 3 rotations of order 2 about axes joining midpoints of opposing edges, and there are 8 rotations of order 3 about axes through a vertex and the center of the opposing face. Since faces sharing an edge through which an axis passes must be the same in a fixed coloring, the order 2 rotations each have 4 fixed colorings. Similarly each of the order 3 rotations has 4 fixed colorings. By the Orbit Formula, then, there are $k = (1/12)(1 \cdot 16 + 3 \cdot 4 + 8 \cdot 4) = 5$ orbits, hence 5 distinguishable colorings. They are of course *rrrr*, *rrrb*, *rrbb*, *rbbb*, and *bbbb*.

Exercises

1. How many distinguishable ways are there to color a regular tetrahedron if each face is red, blue, or green? List them all.

2. How many distinguishable ways are there to color the edges of a square with 2 colors? With 3 colors?

For a rather different sort of application, we may use the Orbit Formula to calculate class numbers of certain groups – the class number $c(G)$ is simply the number of distinct conjugacy classes in G.

Suppose that $H \triangleleft G$ and that $[G:H] = p$, a prime. Assume that $c(H) = d$ and that the conjugacy classes of H are labeled as $C_1, C_2, \ldots, C_d$. Note that G acts by conjugation on the set $S = \{C_1, C_2, \ldots, C_d\}$ – to see why suppose that $C_i = \{y^z : z \in H\}$, with $y \in H$. Take $x \in G$, then $C_i^x = \{y^{zx} : z \in H\} = \{(y^x)^{z^x} : z \in H\}$, another H-conjugacy class since $y^x \in H$ and z^x ranges over H as z does. Also H is in the kernel of that action, so in fact G/H acts on S. Since G/H is cyclic of order p every $x \in G \setminus P$ determines the same orbits in S. We may assume the C_i's labeled so that $C_i^x = C_i$, $1 \le i \le k$ for all $x \in G \setminus H$, and the remaining C_j's lie in G-orbits each of size p. Clearly $C_1, \ldots, C_k$ are each also G-conjugacy classes, and the unions of each of the size p orbits are also G-conjugacy classes. Thus we see that the total number of G-conjugacy classes lying in H is $k + (d - k)/p$.

Proposition 3.1.4 (Burnside, [10]) *Suppose that G and H are as above. Then*

$$c(G) = pk + \frac{c(H) - k}{p},$$

where k is the number of H-classes that are also G-classes.

Proof It has already been observed that there are $k + (d - k)/p$ distinct G-classes in H, $d = c(H)$; we proceed by considering classes not in H.

If $x \in G \setminus H$ and $h, z \in H$, then $(hx)^z = h^z \cdot [z, x^{-1}] \cdot x \in Hx$ since $G' \le H$. Thus H acts by conjugation on the coset Hx. In fact, each H-orbit in Hx is a G-conjugacy class. To see that take a particular element hx in some H-orbit. Since G/H is cyclic of prime order it will suffice to show that the conjugate of hx by any one element of $G \setminus H$ lies in the same H-orbit as hx, so let us conjugate by x^{-1}. But $(hx)^{x^{-1}} = (hx)^h$, in the same H-orbit as hx.

Thus to complete the proof we need to count the H-orbits in each of the $p-1$ cosets of the form Hx, $x \notin H$. To that end we need to calculate numbers of fixed points of elements of H. Consider first an element a lying in one of the first k conjugacy classes $C_1, \ldots, C_k$ of H, which are also classes of G. Thus $^x a = a^b$ for some $b \in H$, and $(hx)^a = hx$, a fixed point, if and only if $h = a^{-1}h \cdot {}^x a = a^{-1}ha^b$, if and only if $a^h = a^b$. That holds if and only if $hb^{-1} \in C_H(a)$, or $h \in C_H(a)b$, and so $\theta(a) = |C_G(a)|$.

Next take $a \in C_j$, $k + 1 \le j \le d$, and recall that $C_j^x \ne C_j$. Let us show that a has no fixed points in Hx. If $(hx)^a = hx$, then $a^{x^{-1}} = a^h$, a contradiction since a^h is in C_j but $a^{x^{-1}}$ is not. Thus $\theta(a) = 0$.

We may now apply the Orbit Formula. Choose $a_i \in C_i$, $1 \le i \le k$. The number of G-conjugacy classes in Hx is

$$|H|^{-1} \sum_{i=1}^{k} \{\theta(a): a \in H\} = |H|^{-1} \sum_{i=1}^{k} |C_i||C_H(a_i)| = |H|^{-1} \sum_{i=1}^{k} |H| = k.$$

Thus $c(G) = k + (d-k)/p + (p-1)k = pk + (d-k)/p.$ △

Exercises

1. Suppose that G is metacyclic as in Proposition 2.2.1, with $s = p$, a prime. Determine $c(G)$.

2. Use (1) to determine $c(D_m)$ and $c(Q_m)$.

3. If p is an odd prime there are two nonabelian groups of order p^3, viz.

$$\langle x, y, z \mid x^p = y^p = z^p = 1,\ xy = yxz,\ xz = zx,\ yz = zy \rangle$$

and

$$\langle x, y \mid x^p = y^{p^2} = 1,\ x^{-1}yx = y^{p+1} \rangle.$$

Show that both have class number $p^2 + p - 1$.

The main result of this chapter, the *Redfield-Pólya Theorem*, is in essence a refinement of the Orbit Formula that makes it possible to count orbits in a wide variety of situations. One further bit of technical apparatus, the *cycle index*, will be needed.

3.2　The Cycle Index

If G acts on S and $|S| = n$ then each $x \in G$ corresponds to an element $\sigma \in \text{Sym}(n)$, which is uniquely a product of disjoint cycles. Recall that the *cycle type* $(j_1, j_2, \ldots, j_n)$ of σ is simply a listing of numbers of cycles; σ has j_k k-cycles for each k. We say as well that $x \in G$ has that same cycle type. Note that conjugate elements have the same cycle type. Define the *monomial* of x to be

$$\text{Mon}(x) = t_1^{j_1} t_2^{j_2} \cdots t_n^{j_n},$$

where t_1, t_2, ..., t_n are distinct (commuting) indeterminates. Then define the *cycle index* of the action of G on S to be the polynomial (say over $\mathbb{Q}$)

$$\mathcal{Z} = \mathcal{Z}_G = \mathcal{Z}_{G,S} = |G|^{-1} \sum \{\text{Mon}(x): x \in G\}.$$

Note that if G has conjugacy classes $K_1, \ldots, K_m$, with representative elements $x_i \in K_i$, then

$$\mathcal{Z} = |G|^{-1} \sum_{i=1}^{m} |K_i| \operatorname{Mon}(x_i) = \sum_{i=1}^{m} |C_G(x_i)|^{-1} \operatorname{Mon}(x_i).$$

Let us compute several examples of cycle indices by tabulating the relevant information about conjugacy classes and cycle types.

1. Let $G = S_3$ and $S = \{1, 2, 3\}$.

σ	1	$(1\,2)$	$(1\,2\,3)$
$|\mathrm{cl}(\sigma)|$	1	3	2
type	$(3,0,0)$	$(1,1,0)$	$(0,0,1)$

Thus $\mathcal{Z} = \frac{1}{6}(t_1^3 + 3t_1 t_2 + 2t_3)$.

2. Let $G = A_4$ and $S = \{1, 2, 3, 4\}$.

σ	1	$(1\,2)(3\,4)$	$(1\,2\,3)$	$(1\,3\,2)$
$|\mathrm{cl}(\sigma)|$	1	3	4	4
type	$(4,0,0,0)$	$(0,2,0,0)$	$(1,0,1,0)$	$(1,0,1,0)$

Thus $\mathcal{Z} = \frac{1}{12}(t_1^4 + 3t_2^2 + 8t_1 t_3)$. Note that this is the same as the cycle index of the rotation group $\mathcal{T}$ of a regular tetrahedron acting either on the faces or the vertices of the tetrahedron.

Exercise

Calculate the cycle index for the action of $\mathcal{T}$ on the set of 6 edges of the tetrahedron.

3. Let $G = \mathcal{O}$, the rotation group of the cube. We shall calculate the cycle indices for three different actions of $\mathcal{O}$: (1) on the set F of 6 faces, (2) on the set V of 8 vertices, and (3) on the set E of 12 edges of the cube. Recall that $\mathcal{O} \cong S_4$, so we may list elements of S_4 as representatives of the conjugacy classes. Note that $(1\,2\,3\,4)$ corresponds to a 90° rotation and $(1\,3)(2\,4)$ to its square, whereas $(1\,2)$ corresponds to the other kind of 180° rotation, with axis joining the midpoints of opposite edges.

σ	1	$(1\,2)$	$(1\,2\,3)$
$|\mathrm{cl}(\sigma)|$	1	6	8
F-type	$(6,0,\ldots)$	$(0,3,0,\ldots)$	$(0,0,2,0,\ldots)$
V-type	$(8,0,\ldots)$	$(0,4,0,\ldots)$	$(2,0,2,0,\ldots)$
E-type	$(12,0,\ldots)$	$(2,5,0,\ldots)$	$(0,0,4,0,\ldots)$

σ	$(1\,2\,3\,4)$	$(1\,3)(2\,4)$
$|\mathrm{cl}(\sigma)|$	6	3
F-type	$(2,0,0,1,\dots)$	$(2,2,0,\dots)$
V-type	$(0,0,0,2,\dots)$	$(0,4,0,\dots)$
E-type	$(0,0,0,3,\dots)$	$(0,6,0,\dots)$

Thus we have

1. $\mathcal{Z}_{\mathcal{O},F} = \frac{1}{24}(t_1^6 + 6t_2^3 + 8t_3^2 + 6t_1^2 t_4 + 3t_1^2 t_2^2)$,

2. $\mathcal{Z}_{\mathcal{O},V} = \frac{1}{24}(t_1^8 + 9t_2^4 + 8t_1^2 t_3^2 + 6t_4^2)$, and

3. $\mathcal{Z}_{\mathcal{O},E} = \frac{1}{24}(t_1^{12} + 6t_1^2 t_2^5 + 8t_3^4 + 6t_4^3 + 3t_2^6)$.

Exercise

Calculate the cycle index for the action of the rotation group $\mathcal{I}$ on the set of 12 vertices of the icosahedron (equivalently on the set of 12 faces of the dodecahedron).

4. Let $G = C_n = \langle \sigma | \sigma^n = 1 \rangle$, acting as a plane rotation group on the set S of n vertices of a regular n-gon. If we label the vertices consecutively as $1, 2, \dots, n$ then we may take σ to be the n-cycle $(1\,2\,\cdots\,n)$. There is a unique (cyclic) subgroup of order d for each positive divisor d of n, and a cyclic group of order d has $\varphi(d)$ generators. Furthermore each $\sigma^k \in C_n$ has order $n/(n,k)$ and is a product of (n,k) disjoint $n/(n,k)$-cycles. Thus if $1 \le d \le n$ and $d \mid n$ there are $\varphi(d)$ elements of order d, each with cycle type $(0, \dots, n/d, \dots, 0)$. As a result

$$\mathcal{Z} = \frac{1}{n}\sum \{\varphi(d) t_d^{n/d} : 1 \le d \le n,\ d \mid n\}.$$

For example, if $n = 18$ then $\mathcal{Z} = \frac{1}{18}(t_1^{18} + t_2^9 + 2t_3^6 + 2t_6^3 + 6t_9^2 + 6t_{18})$.

5. Let $G = D_n$, dihedral of order $2n$, acting on the set of vertices of a regular n-gon. In the presentation

$$D_n = \langle \sigma, \tau | \sigma^n = \tau^2 = (\sigma\tau)^2 = 1 \rangle$$

we may assume that σ is a generating rotation as in the example above, so the cycle structure of each σ^k is already determined. Write $A = \langle \sigma \rangle$.

Let τ be a reflection whose mirror passes through one of the vertices of the n-gon. Since $\sigma^{-k}\tau\sigma^k = \sigma^{-2k}\tau$ we have $\mathrm{cl}(\tau) = \{\sigma^{2k}\tau : k \in \mathbb{Z}\}$. If n is odd then $\langle \sigma^2 \rangle = \langle \sigma \rangle$, so $\mathrm{cl}(\tau)$ is the entire coset $A\tau$ in that case. Each reflection fixes the vertex on its mirror and exchanges the remaining vertices in pairs, so the cycle type is $(1, (n-1)/2, 0, \dots)$.

If n is even there are two classes of reflections, viz. $\mathrm{cl}(\tau)$ and $\mathrm{cl}(\sigma\tau)$, each of size $n/2$. Reflections in $\mathrm{cl}(\tau)$ have mirrors through 2 opposite vertices and cycle type $(2, (n-2)/2, 0, \ldots)$. Those in $\mathrm{cl}(\sigma\tau)$ have mirrors passing through midpoints of opposite edges and cycle type $(0, n/2, 0, \ldots)$.

If n is odd then

$$\mathcal{Z} = \frac{1}{2n} \left[\sum \{\varphi(d) t_d^{n/d} : 1 \le d \le n, \ d \,|\, n\} + n t_1 t_2^{(n-1)/2} \right];$$

if n is even then

$$\mathcal{Z} = \frac{1}{2n} \left[\sum \{\varphi(d) t_d^{n/d} : 1 \le d \le n, \ d \,|\, n\} + \frac{n}{2} (t_1^2 t_2^{(n-2)/2} + t_2^{n/2}) \right].$$

Exercise

If $\sigma \in S_n$ has cycle type $(j_1, \ldots, j_n)$ show that

$$|\mathrm{cl}(\sigma)| = \frac{n!}{\prod_{i=1}^{n} (i^{j_i} \cdot j_i!)}.$$

(This is perhaps best done by "overcounting"; i.e. fill in the slots in all $n!$ ways and then determine how many times each element has been counted and re-counted.)

6. Let $G = S_n$, the symmetric group. Use the abbreviation (j) for the cycle type $(j_1, \ldots, j_n)$, and write $\mathcal{P}$ for the set of all possible cycle types (equivalently the set of all partitions of n). It is immediate from the exercise above that the cycle index for S_n is

$$\mathcal{Z} = \frac{1}{n!} \sum_{(j) \in \mathcal{P}} \frac{n!}{\prod_{i=1}^{n} (i^{j_i} \cdot j_i!)} t_i^{j_i} = \sum_{(j) \in \mathcal{P}} \prod_{i=1}^{n} \frac{1}{(j_i!)} \left(\frac{t_i}{i} \right)^{j_i}.$$

As particular examples we have

$$\mathcal{Z}_{S_4} = \frac{1}{24} (t_1^4 + 6t_2^2 t_2 + 3t_2^2 + 8t_1 t_3 + 6t_4)$$

and

$$\mathcal{Z}_{S_5} = \frac{1}{120} (t_1^5 + 10t_1^3 t_2 + 15t_1 t_2^2 + 20t_1^2 t_3 + 20t_2 t_3 + 30t_1 t_4 + 24t_5).$$

3.3   Enumeration

In the discussion above of colorings of a tetrahedron we passed from a permutation action on the set of 4 faces to the set of $2^4 = 16$ colorings of the
faces. If the faces are labeled 1, 2, 3, and 4 write $D = \{1, 2, 3, 4\}$ for the set
of faces. Let $R = \{r, b\}$, the set of colors. Then the set of colorings can be
identified with the set of all functions $f \colon D \to R$, since a coloring is simply an
assignment of a color to each face.

   In general, suppose that D (the *domain*) and R (the *range*) are 2 finite sets,
and write R^D for the set of all functions from D to R. Note that $|R^D| = |R|^{|D|}$.
If G acts on D, then G also acts on R^D if we define $^\sigma f(d) = f(d^\sigma)$ for all
$f \in R^D$, $\sigma \in G$, and $d \in D$.

   The G-orbits in the set R^D are commonly called *patterns*.

   Suppose that S is a commutative ring with 1 containing the rational field
$\mathbb{Q}$ as a subring, with $1_S = 1_\mathbb{Q}$ (S will often be a polynomial ring, for example).
A function $\mathbf{w} \colon R \to S$ is called a *weight assignment* on R. For each $r \in R$
the ring element $\mathbf{w}(r)$ is called the *weight* of r. A weight assignment $\mathbf{w}$ on R
induces a weight assignment W on R^D via

$$W(f) = \prod \{\mathbf{w}(f(d)) \colon d \in D\}$$

for all $f \in R^D$.

   Suppose that $\mathbf{w}$ is a weight assignment on R and that f_1, $f_2 \in R^D$ are in
the same pattern, say with $f_2 = {}^\sigma f_1$, $\sigma \in G$. Then

$$
\begin{aligned}
W(f_2) &= \prod \{\mathbf{w}(f_2(d)) \colon d \in D\} = \prod \{\mathbf{w}(^\sigma f_1(d)) \colon d \in D\} \\
&= \prod \{\mathbf{w}(f_1(d^\sigma)) \colon d \in D\} = W(f_1)
\end{aligned}
$$

since σ permutes the elements of D. Thus the induced weight assignment is
constant on the patterns in R^D. If F is a pattern define its weight $W(F)$ to
be $W(f)$ for any $f \in F$.

   If U is any subset of R define the *inventory* of U to be

$$\mathrm{Inv}(U) = \sum \{\mathbf{w}(r) \colon r \in U\},$$

with the convention that $\mathrm{Inv}(\emptyset) = 0$. Similarly if $T \subseteq R^D$ define its inventory
to be $\mathrm{Inv}(T) = \sum \{W(f) \colon f \in T\}$.

   In the example above (coloring the tetrahedron) we had domain $D = \{1, 2, 3, 4\}$ and range $R = \{r, b\}$. Take $S = \mathbb{Q}[t_1, t_2]$, polynomials in 2 indeterminates, and define a weight assignment $\mathbf{w} \colon R \to S$ via $\mathbf{w}(r) = t_1$, $\mathbf{w}(b) = t_2$.
As we saw, there are 5 patterns $F_1, \ldots, F_5$, represented by $f_1 = rrrr \in F_1$,
$f_2 = rrrb \in F_2$, $f_3 = rrbb \in F_3$, $f_4 = rbbb \in F_4$, and $f_5 = bbbb \in F_5$. It
is easy to check that $|F_1| = |F_5| = 1$, $|F_2| = |F_4| = 4$, and $|F_3| = 6$, and it

is clear that $W(F_1) = t_1^4$, $W(F_2) = t_1^3 t_2$, $W(F_3) = t_1^2 t_2^2$, $W(F_4) = t_1 t_2^3$, and $W(F_5) = t_2^4$. Consequently

$$\begin{aligned} \mathrm{Inv}(R^D) &= t_1^4 + 4t_1^3 t_2 + 6t_1^2 t_2^2 + 4t_1 t_2^3 + t_2^4 \\ &= (t_1 + t_2)^4 = [\mathrm{Inv}(R)]^{|D|}. \end{aligned}$$

The example illustrates a special case of the next result.

Proposition 3.3.1 *Suppose that R and D are finite sets and that $\mathbf{w} \colon R \to S$ is a weight assignment on R. Suppose that D is partitioned as a union of disjoint subsets $D_1, \ldots, D_m$, and denote by T the set of functions $f \in R^D$ that are constant on each of the subsets D_i. Then*

$$\mathrm{Inv}(T) = \prod_{i=1}^{m} \sum \{\mathbf{w}(r)^{|D_i|} \colon r \in R\}.$$

Proof Define $\psi \colon D \to \{1, 2, \ldots, m\}$ by agreeing that $\psi(d) = j$ if and only if $d \in D_j$. For each $f \in T$ define $\phi = \phi_f \colon \{1, \ldots, m\} \to R$ via $\phi(j) = f(d)$ if and only if $d \in D_j$. Clearly then each $f \in T$ factors as $f = \phi_f \psi$. If $R = \{r_1, \ldots, r_k\}$ then the right hand side of the equation to be proved is

$$\prod_{i=1}^{m} \{\mathbf{w}(r_1)^{|D_i|} + \cdots + \mathbf{w}(r_k)^{|D_i|}\}.$$

To multiply we choose one term from each factor in all possible ways, multiply the terms together, and add the resulting products. To choose a term from each factor is to choose a function $\phi \colon \{1, \ldots, m\} \to R$, then multiplying yields $\prod_{i=1}^{m} \mathbf{w}(\phi(i))^{|D_i|}$, after which we add the results for all such ϕ. But

$$\mathbf{w}(\phi(i))^{|D_i|} = \prod \{\mathbf{w}(\phi(\psi(d))) \colon d \in D_i\} = \prod \{\mathbf{w}(f(d)) \colon d \in D_i\},$$

where $f = \phi\psi \in T$. The choice of all possible ϕ results in all $f \in T$, so the effect of adding is

$$\sum_{f \in T} \prod_{i=1}^{m} \prod_{d \in D_i} \mathbf{w}(f(d)) = \sum_{f \in T} W(f) = \mathrm{Inv}(T).$$

$\triangle$

Corollary 3.3.2 $\mathrm{Inv}(R^D) = [\mathrm{Inv}(R)]^{|D|}$.

Proof If $D = \{d_1, \ldots, d_n\}$ take $D_i = \{d_i\}$, $1 \le i \le n$. $\triangle$

Suppose now that D and R are finite sets as usual, $\mathbf{w} \colon R \to S$ is a weight assignment, and G is a finite group acting on D, hence also on R^D. Recall that if $f \in R^D$ then the weight of f is $\prod\{\mathbf{w}(f(d)): d \in D\}$, and if F is a pattern (i.e. A G-orbit in R^D) then its weight is $W(F) = W(f)$ for any $f \in F$. Define the *pattern inventory* of G to be

$$\text{PI} = \sum\{W(F): \text{all patterns } F \subseteq R^D\}.$$

Suppose, for example, that we define $\mathbf{w}(r) = 1 \in \mathbb{Q}$ for all $r \in R$. Then $W(f) = W(F) = 1$ for all $f \in R^D$ and all patterns F, and so the pattern inventory PI is just the number of patterns. Thus PI, if it can be computed, contains at least as much information as is afforded by the Orbit Formula (Proposition 3.1.2) applied to the action of G on R^F.

Theorem 3.3.3 (Redfield-Pólya) *Suppose that G is a finite group acting on a finite set D, $|D| = k$, with cycle index $\mathcal{Z}$, and that R is a finite set with a weight assignment $\mathbf{w}: R \to S$. Then the pattern inventory for the action of G on R^D is*

$$\text{PI} = \mathcal{Z}\left(\sum_{r \in R} \mathbf{w}(r), \sum_{r \in R} \mathbf{w}(r)^2, \ldots, \sum_{r \in R} \mathbf{w}(r)^k\right).$$

Proof Let $W_1, W_2, \ldots$ be the distinct weights of patterns in R^D, and set

$$\mathcal{F}_i = \{f \in R^D : W(f) = W_i\},$$

$i = 1, 2, \ldots$. Clearly $\mathcal{F}_i$ is a union of patterns; let m_i denote the number of distinct patterns in $\mathcal{F}_i$. Thus $\text{PI} = \sum_i m_i W_i$. Note that G acts on each $\mathcal{F}_i$, and that the G-orbits in $\mathcal{F}_i$ are just the m_i patterns whose union is $\mathcal{F}_i$. If we write θ_i for the character of the action of G on $\mathcal{F}_i$ then, by the Orbit Formula, $m_i = |G|^{-1} \sum\{\theta_i(\sigma): \sigma \in G\}$. Consequently

$$\text{PI} = \sum_i m_i W_i = |G|^{-1} \sum\left\{\sum_i \theta_i(\sigma)W_i: \sigma \in G\right\}.$$

Fix $\sigma \in G$. Then $\sum_i \theta_i(\sigma)W_i$ is the inventory of $\{f \in R^D: {}^\sigma f = f\}$, since $\theta_i(\sigma)$ is just the number of $f \in \mathcal{F}_i$ fixed by σ, and so

$$\text{PI} = |G|^{-1} \sum_{\sigma \in G} \left(\{W(f): f \in R^D \text{ and } {}^\sigma f = f\}\right).$$

If ${}^\sigma f = f$ and $d \in D$ then

$$f(d) = {}^\sigma f(d) = f(d^\sigma) = {}^\sigma f(d^\sigma) = f(d^{\sigma^2}) = \cdots,$$

and f is constant on the σ-cycles in D. Conversely, if $f \in R^D$ is constant on the σ-cycles in D then ${}^\sigma f = f$ (why?). Thus, if the σ-cycles partition D into disjoint sets $D_1, D_2, \ldots$ then $\{f: {}^\sigma f = f\}$ is just the set of all $f \in R^D$ that are constant on each of the sets D_i, i.e. the set T of Proposition 3.3.1. By that proposition

$$\sum \{W(f): {}^\sigma f = f\} = \prod_i \left(\sum \{\mathbf{w}(r)^{|D_i|}: r \in R\} \right)$$

for each fixed $\sigma \in G$. If σ has cycle type $(j_1, \ldots, j_k)$ then among the sets D_i there are j_1 of them for which $|D_i| = 1$, j_2 for which $|D_i| = 2$, etc. Thus

$$\prod_i \left(\sum \{\mathbf{w}(r)^{|D_i|}: r \in R\} \right) = \left(\sum_{r \in R} \mathbf{w}(r) \right)^{j_1} \cdots \left(\sum_{r \in R} \mathbf{w}(r)^k \right)^{j_k}$$

for each σ, and consequently

$$
\begin{aligned}
\text{PI} &= |G|^{-1} \sum_{\sigma \in G} \left(\sum_{r \in R} \mathbf{w}(r) \right)^{j_1(\sigma)} \left(\sum_{r \in R} \mathbf{w}(r)^2 \right)^{j_2(\sigma)} \cdots \left(\sum_{r \in R} \mathbf{w}(r)^k \right)^{j_k(\sigma)} \\
&= \mathcal{Z} \left(\sum_{r \in R} \mathbf{w}(r), \sum_{r \in R} \mathbf{w}(r)^2, \ldots, \sum_{r \in R} \mathbf{w}(r)^k \right).
\end{aligned}
$$

$\triangle$

Corollary 3.3.4 *The total number of different patterns in R^D is* $\mathcal{Z}(|R|, |R|, \ldots, |R|)$.

Proof Set $\mathbf{w}(r) = 1$, all $r \in R$. $\triangle$

The theorem was first proved by Redfield [52] in 1927, but it promptly fell into obscurity. It was rediscovered by Pólya ([51]) ten years later. N. de Bruijn wrote an excellent survey article [19], still very much worth reading, in 1964. The theorem has been used extensively for enumerating graphs with various conditions imposed on them; see [35].

For an example take $D = F$, the set of 6 faces of a cube, and $R = \{\text{red}, \text{blue}, \text{yellow}\}$. Let $S = \mathbb{Q}[r, b, y]$, where r, b, and y are distinct indeterminates, and assign weights by means of $\mathbf{w}(\text{red}) = r$, $\mathbf{w}(\text{blue}) = b$, and $\mathbf{w}(\text{yellow}) = y$. Recall from page 65 that the cycle index $\mathcal{Z}_{O,F}$ is

$$\mathcal{Z} = \frac{1}{24}(t_1^6 + 6t_2^3 + 8t_3^2 + 6t_1^2 t_4 + 3t_1^2 t_2^2).$$

By the Redfield-Pólya Theorem

$$
\begin{aligned}
\text{PI} &= \frac{1}{24}[(r + b + y)^6 + 6(r^2 + b^2 + y^2)^3 + 8(r^3 + b^3 + y^3)^2 \\
&\quad + 6(r + b + y)^2(r^4 + b^4 + y^4) + 3(r + b + y)^2(r^2 + b^2 + y^2)^2] \\
&= r^6 + b^6 + y^6 + r^5 b + r^5 y + r b^5 + r y^5 + b^5 y + b y^5
\end{aligned}
$$

$$+2(r^3b^3 + r^3y^3 + b^3y^3 + r^4b^2 + r^4y^2 + b^4y^2$$

$$+b^2y^4 + r^2b^4 + r^2y^4 + r^4by + rb^4y + rby^4)$$

$$+3(r^3b^2y + r^3by^2 + r^2b^3y + r^2by^3 + rb^3y^2 + rb^2y^3) + 6r^2b^2y^2.$$

(Verify.) The total number of distinguishable colorings is obtained by setting $r = b = y = 1$; there are 57 of them. Note also, though, that the monomial summands of the polynomial PI give the numbers of patterns of various types. For example the monomial r^6 represents the unique pattern with all faces red; the monomial $3r^3b^2y$ represents 3 distinct patterns each having 3 red faces, 2 blue faces, and 1 yellow face. The 2 blue faces can be either adjacent or opposite, and when they are adjacent the yellow face can be adjacent either to both of them or to only one of them.

For a second example let us determine the number of distinguishable necklaces that can be made by using 9 red or blue beads. We assume that beads of the same color are indistinguishable, and that 2 necklaces are indistinguishable if one can be made to look like the other by rotating it and/or turning it over. In effect, then, we imagine each necklace laid out so the beads are at the vertices of a regular 9-gon, which is acted on by the dihedral group D_9.

Take D to be the set of 9 beads, $R = \{\text{red}, \text{blue}\}$, $S = \mathbb{Q}[r, b]$ as above, and $\mathbf{w}(\text{red}) = r$, $\mathbf{w}(\text{blue}) = b$. The cycle index for D_9 is

$$\mathcal{Z} = \frac{1}{18}(t_1^9 + 2t_3^3 + 6t_9 + 9t_1t_2^4),$$

so the pattern inventory is

$$\begin{aligned}
PI &= \frac{1}{18}[(r + b)^9 + 2(r^3 + b^3)^3 + 6(r^9 + b^9) + 9(r + b)(r^2 + b^2)^4] \\
&= r^9 + r^8b + 4r^7b^2 + 7r^6b^3 + 10r^5b^4 \\
&\quad +10r^4b^5 + 7r^3b^6 + 4r^2b^5 + rb^8 + b^9.
\end{aligned}$$

Set $r = b = 1$ to see that there are 46 distinguishable necklaces, and observe, for example, that there are 7 patterns each using 3 red beads and 6 blue beads.

Exercises

1. Let $p \in \mathbb{N}$ be a prime. Discuss the possible necklaces with p beads (a) if red and blue beads are used, and (b) if red, blue, and green beads are used. In particular, determine the number of distinguishable necklaces.

2. In how many distinguishable ways can the faces of a regular tetrahedron be colored if there are n colors available?

3.4 Generating Functions

A different sort of weight assignment is often useful in combinatorics. Suppose that D, R, and G are as usual, let x be an indeterminate, and set $S = \mathbb{Q}[x]$. Let $k: R \to \mathbb{N}$ be a function, and call $k(r)$ the *content* of $r \in R$. For $f \in R^D$ define the content of f to be $K(f) = \sum\{k(f(d)): d \in D\}$. Assign weights on R by means of $\mathbf{w}(r) = x^{k(r)}$, all $r \in R$. Note that then, for $f \in R^D$,

$$W(f) = \prod\{\mathbf{w}(f(d)): d \in D\} = \prod\{x^{k(f(d)):d \in D}\} = x^{\sum_d k(f(d))} = x^{K(f)}.$$

For each $m \in \mathbb{N}$ write $a_m = |\{r \in R: k(r) = m\}|$, and define the *generating function* for R (and k) to be

$$a(x) = \sum\{a_m x^m : m \in \mathbb{N}\},$$

a polynomial whose coefficients count the numbers of elements in R having various contents.

If f_1 and f_2 are in the same pattern F in R^D it is easy to verify that $K(f_1) = K(f_2)$, so we may define $K(F) = K(f)$ for any $f \in F$. If $m \in \mathbb{N}$ set

$$A_m = |\{F: F \text{ is a pattern and } K(F) = m\}|,$$

and define the *pattern counting function* to be

$$A(x) = \sum_m A_m x^m \in \mathbb{Z}[x].$$

Theorem 3.4.1 *Suppose that R and D are finite sets and G is a finite group acting on D, with cycle index $\mathcal{Z}$, generating function $a(x)$, and pattern counting function $A(x)$. Then*

$$A(x) = \mathcal{Z}(a(x), a(x^2), \ldots, a(x^k)).$$

Proof Since $\text{PI} = \sum\{W(F): \text{all patterns } F\}$, and $W(F) = x^{K(F)}$, we have

$$\text{PI} = A_0 + A_1 x + \cdots = A(x).$$

But by the Redfield-Pólya Theorem

$$\text{PI} \quad = \quad \mathcal{Z}\left(\sum_r \mathbf{w}(r), \sum_r \mathbf{w}(r^2), \ldots\right)$$

$$= \quad \mathcal{Z}(a_0 + a_1 x + a_2 x^2 + \cdots, a_0 + a_1 x^2 + a_2 x^4 + \cdots, \ldots)$$

$$= \quad \mathcal{Z}(a(x), a(x^2), \ldots, a(x^k)).$$

$$\triangle$$

If $|D| = k$ define $D^{(m)}$, for $0 \le m \le k$, to be the collection of all m-sets in D, i.e. subsets of size m. Thus $|D^{(m)}| = \binom{k}{m}$. If a group G acts on D then it also acts naturally on each $D^{(m)}$: if $\sigma \in G$ then $\{d_1, d_2, \ldots, d_m\}^\sigma = \{d_1^\sigma, d_2^\sigma, \ldots, d_m^\sigma\}$.

Theorem 3.4.2 *Suppose that G acts on D with cycle index $\mathcal{Z}$, and $|D| = k$. If x is an indeterminate and $0 \le m \le k$ then the number of G-orbits in $D^{(m)}$ is the coefficient of x^m in $\mathcal{Z}(1 + x, 1 + x^2, \ldots, 1 + x^k)$.*

Proof Take $R = \{0, 1\}$, define a content $k: R \to \mathbb{Z}$ via $k(0) = 0$, $k(1) = 1$, and hence assign weights via $\mathbf{w}(0) = 1$, $\mathbf{w}(1) = x$. If $f \in R^D$ then, as observed,

$$W(f) = x^{K(f)} = x^{\Sigma_d f(d)}.$$

Thus $W(f) = x^m$ if and only if precisely m elements of D are mapped by f to 1, i.e. if and only if $f^{-1}(1) \in D^{(m)}$. There is thus a 1–1 correspondence between G-orbits of m-sets in $D^{(m)}$ and patterns of content m in R^D. But the number of patterns of content m is the coefficient A_m in the pattern counting function $A(x)$. Since the generating function is $a(x) = 1 + x$ the result follows from Theorem 3.4.1.   △

Corollary 3.4.3 *The number of G-orbits in D is the coefficient of x in*

$$\mathcal{Z}(1 + x, 1 + x^2, \ldots, 1 + x^k).$$

Proof Identify D with $D^{(1)}$.   △

Corollary 3.4.4 *The total number of orbits of G acting simultaneously on all $D^{(m)}$, for $0 \le m \le k$, is*

$$A_0 + A_1 + \cdots + A_k = \mathcal{Z}(1 + 1, 1 + 1^2, \ldots) = \mathcal{Z}(2, 2, \ldots, 2).$$

For an example take $G = \mathcal{O}$, acting on the set V of 8 vertices of a cube. Recall that the cycle index is $\mathcal{Z} = \frac{1}{24}(t_1^8 + 9t_2^4 + 8t_1^2 t_3^2 + 6t_4^2)$. Thus

$$
\begin{aligned}
A(x) &= \mathcal{Z}(1 + x, 1 + x^2, \ldots, 1 + x^8) \\[2mm]
&= \frac{1}{24}[(1 + x)^8 + 9(1 + x^2)^4 + 8(1 + x)^2(1 + x^3)^2 + 6(1 + x^4)^2] \\[2mm]
&= 1 + x + 3x^2 + 3x^3 + 7x^4 + 3x^5 + 3x^6 + x^7 + x^8.
\end{aligned}
$$

(Verify.) Thus, for example, there are 3 orbits of 2-sets: one orbit, of size 12, consists of pairs of vertices that share an edge; another, also of size 12, consists of pairs of vertices at opposite corners of faces; and the third orbit, of size 4, consists of pairs of vertices at opposite corners of the cube.

Exercises

1. Describe the 3 orbits of 3-sets and the 7 orbits of 4-sets in the example above.

2. If $\mathcal{O}$ acts on the set E of edges of a cube compute the numbers of orbits of m-sets, $0 \le m \le 12$, and describe the orbits of 2-sets explicitly.

3. Suppose that a hollow regular tetrahedron has thin walls, and both the inside and outside of each face is to be colored red or blue. How many distinguishable colorings are there, and how many of them have 3 or fewer of the faces blue?

3.5 The Petersen Graph

This final section can be viewed as an extended list of exercises.

We begin by constructing a well-known graph Γ, the *Petersen* graph. For the set $\mathcal{V}$ of vertices of Γ take the set of 10 transpositions in the symmetric group S_5; we agree that two vertices are joined by an edge in Γ if and only if they are disjoint as permutations. See the figure below.

An *automorphism* of a graph is any permutation σ of its vertex set $\mathcal{V}$ such that u and v are joined by an edge in Γ if and only if u^σ and v^σ are joined by an edge. Denote by G the group Aut(Γ) of all graph automorphisms of Γ.

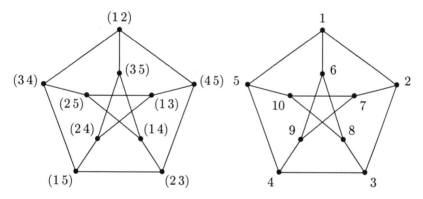

The Petersen Graph

1. Show that S_5 acts faithfully on $\mathcal{V}$ by conjugation, and that each $\sigma \in S_5$ is a graph automorphism. Conclude that S_5 can be viewed as a subgroup of G.

2. Show that G is transitive on $\mathcal{V}$.

3. If H denotes the stabilizer in G of the vertex $(1\,2)$ show that $|H| = 12$. What is H? [*Hint* H includes automorphisms that restrict to give all possible permutations of the three vertices $(3\,4)$, $(3\,5)$, and $(4\,5)$ that are adjacent to $(1\,2)$, and there is a *unique* (nonidentity) $\sigma \in H$ that fixes all of $(3\,4)$, $(3\,5)$, and $(4\,5)$.]

4. Conclude that $|G| = 120$, and hence that $G = S_5$.

5. Relabel the vertices as 1 through 10 (see the figure again), so that $G = S_5$ is now represented as a subgroup of S_{10}. Show that

$$a = (1, 2, 3, 4, 5)(6, 7, 8, 9, 10) \text{ and } b = (3, 7)(8, 9)(4, 10)$$

generate G.

6. Calculate the cycle index $\mathcal{Z}$ of $G \leq S_{10}$. (Suggestion. You may wish to use the GAP command `ConjugacyClasses(G)` to determine representatives and sizes of the conjugacy classes, cycle structure, etc.)

7. We now wish to color the vertices of Γ each red or blue, and two colorings will not be distinguished if there is a graph automorphism carrying one to the other. How many distinguishable colorings are there, and how many of them have half the vertices red, half blue? (Suggestion. You may wish to use a computer algebra package, such as *Maple* [12], to substitute the powers $r^i + b^i$ into the cycle index.)

8. Try to determine the distinguishable colorings having 3 red and 7 blue vertices.

9. Answer questions similar to the previous two questions if three (or more) colors are available.

Chapter 4

Transfer and Splitting

This is the most serious "purely group-theoretical" chapter in the book. We develop some of the techniques that are central to the further study of the structure of finite groups. Substantial applications of some of the results will appear in Chapter 9, where we discuss the structure of Frobenius groups.

4.1 Transfer and Normal Complements

Suppose that G is a group and that $H \leq G$ has finite index n. Let $T = \{t_1, \ldots, t_n\} \subseteq G$ be a (right) transversal for H, and write $x \mapsto \bar{x}$ for the transversal function, i.e. $Hx \cap T = \{\bar{x}\}$ (see Chapter 1).

If A is an abelian group and $\theta: H \to A$ is a homomorphism define a function $V: G \to A$ via $V(x) = \prod_{i=1}^{n} \theta(t_i x \overline{t_i x}^{-1})$. Note that V is well defined since A is abelian; it is called the *transfer* (or *Verlagerung*) from G to A via θ. The transfer was first introduced by I. Schur in 1902.

Proposition 4.1.1 *The transfer $V: G \to A$ is a homomorphism.*

Proof If x, $y \in G$ then

$$
\begin{aligned}
V(xy) &= \prod_i \theta(t_i xy \overline{t_i xy}^{-1}) = \prod_i \theta(t_i x \overline{t_i x}^{-1} \overline{t_i x} y \overline{t_i xy}^{-1}) \\
&= \prod_i \theta(t_i x \overline{t_i x}^{-1}) \cdot \prod_i \theta(\overline{t_i x} y \overline{t_i xy}^{-1}) = V(x)V(y)
\end{aligned}
$$

(we have used Proposition 1.1.1 and the fact that $\{t_1 x, \ldots, t_n x\}$ is also a transversal). $\triangle$

Proposition 4.1.2 *The transfer $V: G \to A$ is independent of the choice of transversal.*

Proof Write $V = V_T$ for the transfer as defined above. Let $S = \{s_1, \ldots, s_n\}$ be another tranversal, denote its transversal function by $x \mapsto \widehat{x}$, and write V_S for the transfer defined in terms of S. We may label the elements of S so that $Hs_i = Ht_i$ for all i, so $\overline{s_i} \doteq t_i$ and $\widehat{t_i} = s_i$. Write $s_i = h_i t_i$, with $h_i \in H$, all i. Now fix $x \in G$. Then $\{\widehat{s_i x} : 1 \le i \le n\} = S$, so there is a permutation $\sigma \in \mathrm{Sym}(n)$ such that $\widehat{s_i x} = s_{i^\sigma}$, all i. Note that $Ht_i x = Hs_i x = H\widehat{s_i x} = Hs_{i^\sigma} = Ht_{i^\sigma}$, so $\overline{t_i x} = t_{i^\sigma}$ as well. Now we calculate:

$$
\begin{aligned}
V_S(x) &= \prod_i \theta(s_i x \widehat{s_i x}^{-1}) = \prod_i \theta(h_i t_i x s_{i^\sigma}^{-1}) = \prod_i \theta(h_i t_i x t_{i^\sigma}^{-1} h_{i^\sigma}^{-1}) \\
&= \prod_i \theta(h_i) \cdot \prod_i \theta(h_{i^\sigma}^{-1}) \cdot \prod_i \theta(t_i x \overline{t_i x}^{-1}) = 1 \cdot V_T(x).
\end{aligned}
$$

$\triangle$

Exercise

Take $G = \mathrm{Sym}(4)$ and let $H = \langle (1\,2\,3\,4), (1\,3) \rangle$, a dihedral subgroup. Take $K = \langle (1\,3)(2\,4) \rangle \triangleleft H$, set $A = H/K$, and let $\theta : H \to A$ be the quotient map. Determine $\ker V$ in this case.

Let us construct a very particular transversal for H in G. Fix $x \in G$ and consider the action of x as a permutation on the set G/H of (right) cosets of H in G. Write the permutation as a product of disjoint cycles (including 1–cycles) as follows:

$$
x \mapsto (Hy_1, Hy_1 x, \ldots, Hy_1 x^{k_1 - 1}) \cdots (Hy_m, Hy_m x, \ldots, Hy_m x^{k_m - 1}).
$$

Then $T = \{t_{ij} = y_i x^j : 1 \le i \le m, 0 \le j \le k_i - 1\}$ is a transversal for H in G. Furthermore the sum of the lengths of the cycles is $n = |G/H|$, i.e. $\sum_{i=1}^m k_i = n$.

In view of Proposition 4.1.2 we may use the transversal T to evaluate the transfer V at x. We first investigate the effect of the transversal function. If $j < k_i - 1$ then $t_{ij} x = y_i x^{j+1} = t_{i(j+1)} \in T$, so $\overline{t_{ij} x} = t_{ij} x$ and $t_{ij} x \overline{t_{ij} x}^{-1} = 1$. However, if $j = k_i - 1$, then $Ht_{ij} x = Hy_i x^{k_i} = Hy_i$, so $\overline{t_{ij} x} = y_i$ and $t_{ij} x \overline{t_{ij} x}^{-1} = y_i x^{k_i} y_i^{-1}$. Note also that $y_i x^{k_i} y_i^{-1} \in H$ for each i.

We have proved the next proposition, which will be used later in an application of the transfer.

Proposition 4.1.3 *Fix $x \in G$. Then there are elements $y_1, \ldots, y_m \in G$ and positive integers $k_1, \ldots, k_m$ such that*

1. $\sum_{i=1}^m k_i = [G : H]$,

2. $y_i x^{k_i} y_i^{-1} \in H$, *all i, and*

3. $V(x) = \prod_{i=1}^m \theta(y_i x^{k_i} y_i^{-1})$.

Exercise

If $H \leq Z(G)$ and $n = [G:H]$ show that $V(x) = \theta(x^n)$, all $x \in G$.

Recall that the derived subgroup G' of G is the smallest normal subgroup of G for which the quotient group is abelian. Thus in particular $G' = \cap\{K: K \triangleleft G \text{ and } G/K \text{ is abelian}\}$.

Let us "localize" the above observation. For any prime p define

$$G'_p = \cap\{K: K \triangleleft G \text{ and } G/K \text{ is an abelian } p\text{-group}\} \triangleleft G.$$

Note that each K in the definition contains G'. Thus $G'_p \geq G'$, and G/G'_p is abelian. The subgroup G'_p is called the *p-commutator subgroup* of G.

Proposition 4.1.4 *For each prime p the p-commutator subgroup G'_p is normal in G, and G/G'_p is an abelian p-group.*

Proof The only part of the proposition requiring proof is that G/G'_p is a p-group. To establish that it will suffice to show that if $H, K \triangleleft G$, with G/H and G/K both p-groups, then $G/(H \cap K)$ is also a p-group. Define $\theta: G/H \times G/K$ via $\theta(x) = (Hx, Kx)$, and note that $\ker \theta = H \cap K$. Thus $G/(H \cap K)$ is isomorphic with a subgroup of the p-group $G/H \times G/K$. $\triangle$

Exercise

Calculate G'_p for all p if (a) $G = \text{Alt}(4)$, (b) $G = \text{Sym}(n)$, and (c) G is the split metacyclic group $\langle a, b | a^{10} = b^6 = 1, b^{-1}ab = a^{-1} \rangle$.

Proposition 4.1.5 *If G is a finite group then $G' = \cap_p G'_p$ and $G/G' \cong \prod_p G/G'_p$.*

Proof Since G/G' is abelian it is the product of its distinct primary subgroups, say $G/G' = H_1/G' \times \cdots \times H_k/G'$, with H_i/G' p_i-primary. Each H_i is normal in G. Fix i and define K_i via $K_i/G' = \prod_{j \neq i} H_j/G'$, so $K_i \triangleleft G$ and $G/G' = H_i/G' \times K_i/G'$. Then $G/K_i \cong G/G'/K_i/G' \cong H_i/G'$, the abelian p_i-group of maximal order in G/G'. Since all abelian images of G are images of G/G' it follows that $K_i = G'_{p_i}$. Now $H_i/G' \cong G/G'_{p_i}$, so $G/G' \cong \prod_p G/G_p$, and also $\cap_i(K_i/G') = G' = 1_{G/G'} = (\cap_p G'_p)/G'$, i.e. $\cap_p G'_p = G'$. $\triangle$

Proposition 4.1.6 *If G is a finite group and $P \in \text{Syl}_p(G)$ then $PG'_p = G$ and $G/G'_p \cong P/(P \cap G'_p)$.*

Proof The index $[G:G'_p]$ is a power of p and a divisor of $|G|$, so if $q \neq p$ is a prime and $q^e \big| |G|$ then $q^e \big| |G'_p|$. Thus $|PG'_p| \big| |G|$ and $G = PG'_p$. Furthermore $G/G'_p = PG'_p/G'_p \cong P/(P \cap G'_p)$ by the standard isomorphism theorem. $\triangle$

Recall that the *commutator* of two group elements x and y is $[x, y] = x^{-1}y^{-1}xy$. If H, $K \leq G$ then $[H, K] = \langle [x, y] : x \in H, \ y \in K \rangle$; in particular $G' = [G, G]$.

If G is a group and $H \leq G$ define the *focal subgroup* of H in G to be

$$\mathrm{Foc}_G(H) = \langle [y, x] : y \in H, x \in G, \ \text{and} \ [y, x] \in H \rangle.$$

Clearly $H' \leq \mathrm{Foc}_G(H) \leq H \cap G'$, and in particular $H/\mathrm{Foc}_G(H)$ is abelian (why is $\mathrm{Foc}_G(H) \triangleleft H$?).

Theorem 4.1.7 *Suppose that* $P \in \mathrm{Syl}_p(G)$. *Then*

$$\mathrm{Foc}_G(P) = P \cap G'_p \ \ \text{and} \ \ P/\mathrm{Foc}_G(P) \cong G/G'_p.$$

Proof It is clear from the definitions that $\mathrm{Foc}_G(P) \leq P \cap G' \leq P \cap G'_p$. For the reverse inclusion set $A = P/\mathrm{Foc}_G(P)$, let $\theta : P \to A$ be the quotient map, and apply the transfer V from G to P via θ. If $x \in P$ then, by Proposition 4.1.3, $V(x) = \prod_{i=1}^{m} \mathrm{Foc}_G(P) y_i x^{k_i} y_i^{-1}$ for appropriate $y_i \in G$, $k_i \in \mathbb{N}$. But $[x^{k_i}, y_i] \in \mathrm{Foc}_G(P)$ since $x^{k_i} \in P$, $y_i x^{k_i} y_i^{-1} \in P$, and so $y_i x^{k_i} y_i^{-1} x^{-k_i} = [x^{k_i}, y_i]^{-1} \in P$. As a result $\mathrm{Foc}_G(P) y_i x^{k_i} y_i^{-1} = \mathrm{Foc}_G(P) x^{k_i}$, and consequently

$$
\begin{aligned}
V(x) &= \prod_i \mathrm{Foc}_G(P) x^{k_i} \\
&= \mathrm{Foc}_G(P) x^{\Sigma_i k_i} = \mathrm{Foc}_G(P) x^{[G:P]}.
\end{aligned}
$$

Since $x \in P$ and $p \nmid [G:P]$, x and $x^{[G:P]}$ have the same order, so in fact $\mathrm{Foc}_G(P) x$ is also in the image of V. Thus V maps P (and *a fortiori* G) *onto* $P/\mathrm{Foc}_G(P)$.

We now have $G/\ker V \cong P/\mathrm{Foc}_G(P)$, an abelian p-group, and so $\ker V \geq G'_p$. Apply Proposition 4.1.6:

$$
\begin{aligned}
[G : \ker V] &= [P : \mathrm{Foc}_G(P)] \geq [P : P \cap G'_p] \\
&= [PG'_p : G'_p] = [G : G'_p] \geq [G : \ker V],
\end{aligned}
$$

and all are equal. The last inequality holds since $G/\ker V$ is an abelian p-group, hence $\ker V \geq G'_p$. For the isomorphism see Proposition 4.1.6. $\triangle$

Proposition 4.1.8 *Suppose that* $P \in \mathrm{Syl}_p(G)$ *and set* $N = N_G(P)$. *If* P *is abelian then* $\mathrm{Foc}_G(P) = \mathrm{Foc}_N(P)$.

Proof Inclusion right-to-left is clear. Let $[y, x]$ be a generator for $\mathrm{Foc}_G(P)$, i.e. $y \in P$, $x \in G$, and $[y, x] = y^{-1}y^x \in P$. Thus also $z = y^x \in P$, and $z \in P^x$, so P and P^x are both Sylow p-subgroups of $C_G(z)$ (P is abelian). By the Sylow Theorem there is some $w \in C_G(z)$ with $P = P^{xw}$, so $xw \in N$. But now $[y, xw] = y^{-1}y^{xw} = y^{-1}z^w = y^{-1}z = y^{-1}y^x = [y, x]$, and $\mathrm{Foc}_G(P) \leq \mathrm{Foc}_N(P)$. $\triangle$

Now an important definition. If G is a group and $H \leq G$, then a *normal complement* for H in G is a normal subgroup K of G satisfying $G = KH$ and $K \cap H = 1$, i.e. $G = K \rtimes H$, a split extension. If G is finite then a normal complement for a Sylow p-subgroup P is called a *normal p-complement*.

For an easy class of examples suppose that G is metacyclic, as in Proposition 2.2.3, with $s = p^e$, p a prime, and $p \nmid m$. Then A is a normal p-complement.

Exercise

If a group G has a normal p-complement K show that K is unique, and hence K char G. Show that $K = \{x \in G : p \nmid |x|\}$.

Theorem 4.1.9 (Burnside) *If $P \in \mathrm{Syl}_p(G)$ and $P \leq Z(N_G(P))$ then G has a normal p-complement K, which is characteristic in G.*

Proof Set $N = N_G(P)$. Note that P is abelian, so $\mathrm{Foc}_G(P) = \mathrm{Foc}_N(P)$ by Proposition 4.1.8. Also $\mathrm{Foc}_N(P) = \langle [y,x] : y \in P, x \in N \rangle = 1$ since $P \leq Z(N)$. Thus $P \cap G'_p = 1$ and $G = PG'_p$ by Proposition 4.1.6 and Theorem 4.1.7, so $K = G'_p$ is a normal p-complement. For K char G see the exercise preceding the theorem. $\triangle$

Note that the hypothesis $P \leq Z(N_G(P))$ in Burnside's Theorem is equivalent with $N_G(P) = C_G(P)$; the theorem is sometimes stated accordingly.

There are numerous applications of Burnside's Theorem; we shall see just a few of them.

Proposition 4.1.10 *If $H \leq G$ then $C_G(H) \triangleleft N_G(H)$ and $N_G(H)/C_G(H)$ is isomorphic with a subgroup of $\mathrm{Aut}(H)$.*

Proof Define a homomorphism θ from $N_G(H)$ to $\mathrm{Aut}\, H$ via $x \mapsto \theta_x$, where $\theta_x(h) = {}^x h$, $h \in H$, and check that $\ker \theta = C_G(H)$. $\triangle$

Theorem 4.1.11 *Suppose that p is the smallest prime divisor of $|G|$, and that $P \in \mathrm{Syl}_p(G)$ is cyclic. Then G has a normal p-complement, and in particular G is not simple.*

Proof If $|P| = p^k$ then $|\mathrm{Aut}\, P| = \varphi(p^k) = p^{k-1}(p-1)$ since P is cyclic, and $P \leq C_G(P)$, so $p \nmid |N_G(P)/C_G(P)|$, and $[N_G(P) : C_G(P)]$ divides $p - 1$ by Proposition 4.1.10. But also $[N_G(P) : C_G(P)] \mid |G|$, whose smallest prime divisor is p, so $[N_G(P) : C_G(P)] = 1$, $N_G(P) = C_G(P)$, and Burnside's Theorem (4.1.9) applies. $\triangle$

Corollary 4.1.12 *If G is simple, p is the smallest prime divisor of $|G|$, and $P \in \mathrm{Syl}_p(G)$, then P is not cyclic.*

Note that the corollary rules out many possible orders for simple groups, viz. all pm, where $p \nmid m$ and all prime divisors of m are larger than p. The first candidate for a Sylow 2-subgroup of a simple group is a Klein 4-group – it occurs in $\mathrm{Alt}(5)$.

If $n \in \mathbb{Z}$ and p is a prime let us write $p^k \, \| \, n$ to indicate that p^k *exactly divides* n, i.e. $p^k \mid n$ but $p^{k+1} \nmid n$.

Proposition 4.1.13 *Suppose that G is simple, p is a prime, and $p \, \| \, |G|$. If $P \in \mathrm{Syl}_p(G)$ then $1 \neq [N_G(P) : C_G(P)] \, \big| \, p - 1$.*

Proof Since $|\operatorname{Aut} P| = p - 1$ the divisibility follows from Proposition 4.1.10. If $N_G(P) = C_G(P)$ then G would not be simple, by Burnside's Theorem. $\triangle$

As a sample application suppose that G is simple and that $|G| = 1900 = 2^2 \cdot 5^2 \cdot 19$. Then $|\mathrm{Syl}_{19}(G)| = 20$. If $P \in \mathrm{Syl}_{19}(G)$ then $[G : N_G(P)] = 20$, so $|N_G(P)| = 5 \cdot 19$. By the proposition $[N_G(P) : C_G(P)] \neq 1$, so $|C_G(P)| = 19$. But then the proposition also tells us that $5 \mid 19 - 1 = 18$, a contradiction.

Exercise

If p and q are primes, say with $p < q$, and if $|G| = p^2 q^2$, show that G is not simple.

One more (direct) application of Burnside's Theorem.

Theorem 4.1.14 *Suppose that G is a nonabelian simple group and p is the smallest prime divisor of $|G|$. If p is odd then $p^3 \, \big| \, |G|$; if $p = 2$ then either $8 \, \big| \, |G|$ or $12 \, \big| \, |G|$.*

Proof Suppose that $p^3 \nmid |G|$, and choose $P \in \mathrm{Syl}_p(G)$; note that P is abelian. In fact, by Corollary 4.1.12 we see that $P \cong \mathbb{Z}_p \oplus \mathbb{Z}_p$, so $\operatorname{Aut} P \cong \mathrm{GL}(2, p)$ and $|\operatorname{Aut} P| = (p-1)^2 p(p+1)$ by Proposition 2.6.1. Write $N = N_G(P)$ and $C = C_G(P)$. By Burnside's Theorem (4.1.9) $[N : C] \neq 1$, and by Proposition 4.1.10 $[N : C] \, \big| \, |\operatorname{Aut} P|$. But also $[N : C] \, \big| \, |G|$, and p is the smallest prime divisor of $|G|$, so $[N : C]$ is prime to $p - 1$. Furthermore $P \leq C$, so $[N : C]$ is also prime to p, hence $[N : C] \, \big| \, p + 1$. But then $[N : C]$ must *equal* $p + 1$, and $p + 1$ must be prime, for otherwise $[N : C]$ and $p + 1$ would share a prime factor smaller than p. That is possible only if $p = 2$ and $p + 1 = 3$; thus 3 and 4 are both divisors of $|G|$. $\triangle$

Burnside conjectured ([10], page 503) that simple groups must have even order. That was proved in 1963 in the monumental paper [24] of Feit and Thompson.

Burnside also noticed that all simple groups known at the time his book was written had orders divisible by 12 ([10], page 330), but he seems not to have made a corresponding conjecture. The first (nonabelian) simple groups with orders prime to 3 (hence divisible by 8) were discovered by Suzuki [61] in 1960.

To conclude this section let us aim at a converse for Proposition 2.2.2, as was promised in Chapter 2.

Proposition 4.1.15 *If all Sylow subgroups of G are cyclic then G is solvable.*

Proof Induction on $|G|$; the result clearly holds for $|G|$ sufficiently small. If p is the smallest prime divisor of $|G|$ and $P \in \mathrm{Syl}_p(G)$ then G has a normal p-complement K by Theorem 4.1.11. By induction K is solvable, and $G/K \cong P$ is solvable, so G is solvable. $\triangle$

Proposition 4.1.16 *Suppose that G is a finite group and that for some $k \geq 2$ the terms $G^{(k-1)} \geq G^{(k)} \geq G^{(k+1)}$ in the derived series have quotients $G^{(k-1)}/G^{(k)}$ and $G^{(k)}/G^{(k+1)}$ that are both cyclic. Then $G^{(k+1)} = G^{(k)}$.*

Proof Set $H = G^{(k-2)}/G^{(k+1)}$, so that $H' = G^{(k-1)}/G^{(k+1)}$, $H^{(2)} = G^{(k)}/G^{(k+1)}$, and $H^{(3)} = 1$. Also $H'/H^{(2)} \cong G^{(k-1)}/G^{(k)}$, so $H'/H^{(2)}$ and $H^{(2)}$ are both cyclic; say $H^{(2)} = \langle x \rangle$. Map H into $\mathrm{Aut}\, H^{(2)}$ by restricting inner automorphisms; the kernel of the map is $C_H(H^{(2)}) = C_H(x)$. Since $H^{(2)}$ is cyclic $\mathrm{Aut}\, H^{(2)}$ is abelian (Theorem 2.1.3), so $H/C_H(x)$ is abelian and hence $H' \leq C_H(x) = C_H(H^{(2)})$, i.e. $H^{(2)} \leq Z(H')$. But then $H'/Z(H')$ is cyclic, being a homomorphic image of $H'/H^{(2)}$. Thus H' is abelian, so $H^{(2)} = 1$, i.e. $G^{(k+1)} = G^{(k)}$. $\triangle$

The parameters m, s, r, and t in the next theorem are as in Proposition 2.2.1.

Theorem 4.1.17 *Suppose that all Sylow subgroups of G are cyclic. Then G is split metacyclic, with $A = G'$ of order m, $s = [G{:}G']$, $t = 0$, and $(m, s(r-1)) = 1$.*

Proof If G is abelian it is cyclic and the conclusion holds trivially, so we may assume it is not abelian. Note that all subgroups and homomorphic images of G have all Sylow subgroups cyclic. In particular $G'/G^{(2)}$ and $G^{(2)}/G^{(3)}$, being abelian, are cyclic. Thus $G^{(3)} = G^{(2)}$ by Proposition 4.1.16, and $G^{(2)} = 1$ by Proposition 4.1.15. Consequently G *is* metacyclic, with $A = G' = \langle a \rangle$, say, of order m, and $G/A = \langle Ab \rangle$ of order s, with $b^{-1}ab = a^r$ and $b^s = a^t$ as in Proposition 2.2.1. Recall (Exercise 4, page 30) that $G' = \langle a^{r-1} \rangle$, so $(m, r-1) = 1$. Also $m \,\big|\, t(r-1)$ and $0 \leq t < m$, so $t = 0$ and G is split. If p is prime and $p \,\big|\, (m, s)$, then the subgroup $\langle a^{m/p}, b^{s/p} \rangle$ has order p^2 and is not cyclic, a contradiction, and the proof is complete. $\triangle$

Corollary 4.1.18 *If G is a group having square-free order then G is split metacyclic with $(m, s(r-1)) = 1$.*

Exercise

Let $G = \langle a, b | a^{14} = b^3 = 1, \ b^{-1}ab = a^9 \rangle$, split metacyclic of order 42. Since $(m, s) = (14, 3) = 1$ all Sylow subgroups are cyclic by Proposition 2.2.2. Conclude from Theorem 4.1.17 that $(m, r-1) = 1$. But $(m, r-1) = (14, 8) = 2$, so $2 = 1$??? What went wrong?

4.2   Hall Subgroups

A subgroup H of a finite group G is called a *Hall subgroup* if $|H|$ and $[G:H]$ are relatively prime. For example, if G has all Sylow subgroups cyclic then its derived group G' is a Hall subgroup by Theorem 4.1.17.

For any finite group G write $\pi(G)$ to denote the set of prime divisors of $|G|$.

If π is any set of prime numbers write π' for the complementary set of primes. An integer n is called a π-*number* if all its prime divisors are in π. Say that a group G is a π-*group* if $|G|$ is a π-number. If $H \leq G$ then H is a *Hall* π-subgroup if H is both a Hall subgroup and is a π-group, and in addition $[G:H]$ is a π'-number. Write $\text{Hall}_\pi(G)$ for the set (possibly empty) of all Hall π-subgroups of G.

Subgroups H and K of a group G are called *complements* of each other if $HK = G$ and $H \cap K = 1$ (no assumptions about normality).

Exercises

1. If H and K are complements in G and H is a Hall subgroup show that K is a Hall $\pi(H)'$-subgroup. Conversely, if H and K are Hall subgroups with $\pi(H) \cap \pi(K) = \emptyset$ and $\pi(H) \cup \pi(K) = \pi(G)$, show that they are complements to one another.

2. Suppose that H is a normal Hall subgroup of G and $|H| = n$. Show that $H = \{x \in G : x^n = 1\}$.

Proposition 4.2.1 (The Frattini Argument) *Suppose that $H \triangleleft G$ and $P \in \text{Syl}_p(H)$. Then $G = N_G(P)H$.*

Proof If $x \in G$ then $P^x \leq H^x = H$, so $P^x \in \text{Syl}_p(H)$. Take $h \in H$ with $P^x = P^h$, so $P^{xh^{-1}} = P$, $xh^{-1} \in N_G(P)$, and $x \in N_G(P)h \subseteq N_G(P)H$. $\triangle$

Theorem 4.2.2 (I. Schur) *A normal abelian Hall subgroup A of G has a complement.*

Proof Write $G/A = \{\alpha, \beta, \gamma, \ldots\}$ and choose a transversal $\{t_\alpha : \alpha \in G/A\}$ for A in G, with, of course, $t_\alpha \in \alpha$. Then $At_\alpha t_\beta = \alpha\beta = At_{\alpha\beta}$, and so $t_\alpha t_\beta = f(\alpha, \beta) t_{\alpha\beta}$ for some $f(\alpha, \beta) \in A$, all α, β. It follows (verify) from the associative law that

$$f(\alpha, \beta) f(\alpha\beta, \gamma) = {}^{t_\alpha} f(\beta, \gamma) f(\alpha, \beta\gamma),$$

all α, β, γ (the function $f : G/A \times G/A \to A$ is called a *factor set*, or a *factor system*). Set $n = |A|$ and $m = [G : A]$, so $(n, m) = 1$. For each $\alpha \in G/A$ set $g(\alpha) = \prod \{f(\alpha, \delta) : \delta \in G/A\}$, and observe that

$$\prod_\gamma [f(\alpha, \beta) f(\alpha\beta, \gamma)] = \prod_\gamma [{}^{t_\alpha} f(\beta, \gamma) f(\alpha, \beta\gamma)],$$

$$\left[\prod_\gamma f(\alpha, \beta) \right] \left[\prod_\gamma f(\alpha\beta, \gamma) \right] = {}^{t_\alpha} \left[\prod_\gamma f(\beta, \gamma) \right] \left[\prod_\gamma f(\alpha, \beta\gamma) \right],$$

and so

$$f(\alpha, \beta)^m g(\alpha\beta) = {}^{t_\alpha} g(\beta) g(\alpha)$$

for all β, $\gamma \in G/A$. Choose $k \in \mathbb{Z}$ so that $km \equiv 1 \pmod{n}$, set $h(\delta) = g(\delta)^{-k}$, all $\delta \in G/A$, and take $-k$th powers to see that

$$f(\alpha, \beta)^{-1} h(\alpha\beta) = {}^{t_\alpha} h(\beta) h(\alpha) \qquad (*)$$

for all α, $\beta \in G/A$.

Now set $s_\alpha = h(\alpha) t_\alpha$, all $\alpha \in G/A$, to obtain a new transversal for A in G, and set $B = \{s_\alpha : \alpha \in G/A\}$. If B is a subgroup of G then it is clearly a complement for A.

If s_α, $s_\beta \in B$ then

$$\begin{aligned} s_\alpha s_\beta &= h(\alpha) t_\alpha h(\beta) t_\beta = h(\alpha) \cdot {}^{t_\alpha} h(\beta) t_\alpha t_\beta \\ &= f(\alpha, \beta)^{-1} h(\alpha\beta) t_\alpha t_\beta = h(\alpha\beta) t_{\alpha\beta} = s_{\alpha\beta} \end{aligned}$$

by equation $(*)$ above, and the proof is complete. $\qquad \triangle$

Theorem 4.2.3 (Schur-Zassenhaus) *A normal Hall subgroup H of G has a complement.*

Proof We use induction on $|G|$, the result being trivially true if G is small enough.

Suppose first that there is a prime p dividing $|H|$ and a Sylow p-subgroup P of H that is not normal in G. Then $G \neq N = N_G(P)$. By the Frattini Argument (Proposition 4.2.1) $G = NH$, and $G/H = (NH)/H \cong N/(N \cap H)$, so $[N : N \cap H] = [G : H]$. Thus $N \cap H$ is a normal Hall subgroup of N, and by induction there is a complement K to $N \cap H$ in N. But then $|K| = [N : N \cap H] = [G : H]$, so K is also a complement to H in G.

Thus we may suppose that all Sylow subgroups of H are normal in G. In particular H is nilpotent, and $Z = Z(H) \neq 1$. Since Z char $H \triangleleft G$ we have $Z \triangleleft G$. Now H/Z is a normal Hall subgroup of G/Z, and by induction there is a complement K/Z to H/Z in G/Z. Note that $|K/Z| = [G/Z:H/Z] = [G:H]$. If $K \neq G$ then Z is a Hall subgroup of K (since $Z \leq H$ and $[K:Z] = [G:H]$), and by induction there is a complement L to Z in K. Then $|L| = [K:Z] = [G:H]$, so L is also a complement to H in G.

Finally, suppose that $K = G$. Then $[G:H] = [K:Z] = [G:Z]$, so $H = Z$ is abelian, and H has a complement by Theorem 4.2.2. $\triangle$

Recall that a finite abelian group A is called *elementary abelian* if all nonidentity elements have order p for some prime p. Thus $A \cong \oplus_{i=1}^k \mathbb{Z}_p$ for some k, and A (written additively) is a $\mathbb{Z}_p$-vector space.

Proposition 4.2.4 *Suppose that G is a finite solvable group and that $N \neq 1$ is a minimal normal subgroup of G. Then N is an elementary abelian p-group for some prime p.*

Proof Suppose that $K \neq 1$ is a minimal normal subgroup of N. Then $\langle \cup \{K^x : x \in G\} \rangle$ is contained in N and is normal in G, so it is equal to N. Each K^x is minimal normal in N and two of them, K^x and K^y, are either equal or intersect only in 1. Thus for some subcollection $\{K^{x_1}, \ldots, K^{x_\ell}\}$ we have $N = K^{x_1} \times \cdots \times K^{x_\ell}$. If $L \triangleleft K^{x_i}$ then $L \triangleleft N$. Thus $L = 1$ or K^{x_i}, since K^{x_i} is minimal normal in N; i.e. K^{x_i} is simple. Since it is also solvable we have $K^{x_i} \cong \mathbb{Z}_p$ for some prime p. $\triangle$

Theorem 4.2.5 (H. Zassenhaus) *Suppose that H is a normal Hall subgroup of G, and suppose that either H or G/H is solvable. Then any two complements to H in G are conjugate.*

Proof Let K_1 and K_2 be complements for H; note that they are both transversals for H in G. We consider three cases.

Case 1. Assume that H is abelian. For each $a \in K_2$ there are a unique $x \in K_1$ and $\alpha(x) \in H$ so that $a = \alpha(x)x$, determining a function $\alpha: K_1 \to H$. If also $b \in K_2$, with $b = \alpha(y)y$, then $Hab = HaHb = HxHy = Hxy$, so $ab = \alpha(xy)xy$. On the other hand $ab = \alpha(x)x\alpha(y)y = \alpha(x) \cdot {}^x\alpha(y)xy$, so

$$\alpha(xy) = \alpha(x) \cdot {}^x\alpha(y) \qquad\qquad (*)$$

for all $x, y \in K_1$ (the function α is called a *cocycle*).

Say that $|K_i| = [G:H] = n$. Set $u = \prod\{\alpha(y): y \in K_1\} \in H$. By the equation $(*)$ we have

$$u = \prod_y \alpha(xy) = \prod_y [\alpha(x) \cdot {}^x\alpha(y)] = \alpha(x)^n \cdot {}^x u,$$

all $x \in K_1$. Since H is abelian and $(|H|, n) = 1$ the map $h \mapsto h^n$ is an automorphism of H. Choose $v \in H$ so that $v^n = u$; then $v = \alpha(x) \cdot {}^x v$ for all $x \in K_1$. But now $^v x = vxv^{-1}x^{-1}x = v(^x v)^{-1}x = \alpha(x)x = a \in K_2$, all $x \in K_1$, and $^v K_1 = K_2$.

Case 2. Assume that H is solvable and use induction on $|G|$. Suppose first that there exists $N \triangleleft G$ with $1 \neq N < H$. Then $K_1 N/N$ and $K_2 N/N$ are conjugate in G/N by induction, since they are complements to H/N. Thus $\exists x \in G$ such that $^x K_1 \leq K_2 N$. But then $^x K_1$ and K_2 are complements of N in $K_2 N < G$, and N is solvable. By induction $\exists y \in K_2 N$ with $^{yx} K_1 = K_2$, as desired.

Suppose then that no such N exists. Then H is solvable and has no nontrivial characteristic subgroups; in particular $H' = 1$, so H is abelian and we are back in Case 1.

Case 3. Assume that G/H is solvable.

Some preparation will be useful. If $x \in G$ then $x = hk_1$, uniquely, with $h \in H$, $k_1 \in K_1$, and the map $\varphi_1 \colon x \mapsto k_1$ is a homomorphism from G onto K_1 with kernel H. It induces an isomorphism $\theta_1 \colon G/H \to K_1$. Similarly there is an isomorphism $\theta_2 \colon G/H \to K_2$, and so $\theta = \theta_2 \theta_1^{-1}$ is an isomorphism from K_1 to K_2. Note that θ provides a 1–1 correspondence between subgroups L_1 of K_1 and L_2 of K_2 as follows: $\theta \colon L_1 \leftrightarrow L_2 = L_1 H \cap K_2$, and $L_1 \triangleleft K_1$ if and only if $L_2 \triangleleft K_2$.

Now the proof. Once again use induction on $|G|$. Since $K_1 \cong G/H$ it is solvable, and a minimal normal subgroup of K_1 is an elementary abelian p-group P_1 for some prime p, by Proposition 4.2.4. Set $P_2 = P_1 H \cap K_2 \cong P_1$, minimal normal in K_2 by the discussion above. Note that both P_1 and P_2 are in $\mathrm{Syl}_p(P_1 H)$, so $\exists x \in P_1 H$ with $P_2^x = P_1$. Set $N = N_G(P_1)$. Since $P_1 \triangleleft K_1$ and $P_1 = P_2^x \triangleleft K_2^x$ we have $K_1, K_2^x \leq N$. Thus $N = G \cap N = (HK_1) \cap N = (H \cap N)K_1$ (since $K_1 \leq N$), and similarly $N = (H \cap N)K_2^x$, i.e. K_1 and K_2^x are complements to $H \cap N$ in N. Now K_1/P_1 and K_2^x/P_1 are complements to $(H \cap N)P_1/P_1$ in N/P_1, and by induction $\exists y \in N$ so that $K_1/P_1 = K_2^{xy}/P_1$, which says that $K_2^{xy} \leq K_1 P_1 = K_1$, and finally $K_2^{xy} = K_1$. $\triangle$

It must be noted that at least one of $|H|$ and $|G/H|$ is an odd number in Theorem 4.2.5, since H is a Hall subgroup, and consequently at least one of H and G/H must be solvable by the Feit-Thompson Theorem mentioned on page 82. Thus the conclusion to the theorem holds under *all* circumstances.

It should perhaps be mentioned that the proofs of Theorems 4.2.2 and 4.2.5, in utilizing factor sets and cocycles, represented brief excursions into the subject of *cohomology of groups*, a subject of considerable importance in its own right, especially for the general study of group extensions. See e.g. [33] or [37] for an introductory account.

As an application of the Schur-Zassenhaus Theorem (4.2.3) we prove next a remarkable generalization, for solvable groups, of the Sylow Theorems, due to Philip Hall.

Theorem 4.2.6 (P. Hall, 1928) *Suppose that G is a finite solvable group and that π is a set of primes. Then G has a Hall π-subgroup, and any two* *of them are conjugate. Furthermore every π-subgroup of G is contained in a Hall π-subgroup.*

Proof If K is a π-subgroup of G let us use induction on $|G|$ to show that $K \leq H$ for some $H \in \mathrm{Hall}_\pi(G)$. Take N minimal normal in G; N is an elementary abelian p-group for some prime p by Proposition 4.2.4. Then KN/N is a π-subgroup of G/N; by induction there is a Hall π-subgroup M/N of G/N with $K \leq M$. Now $[M:N]$ is a π-number and $[G:M]$ is a π'-number, so if $p \in \pi$ then $M \in \mathrm{Hall}_\pi(G)$, and the proof is complete with $H = M$. Suppose then that $p \in \pi'$, so N is a normal Hall subgroup of M, and N has a complement L in M by Theorem 4.2.3 (or 4.2.2 for that matter). Now $|L| = [M:N]$ is a π-number, and $[G:L] = [G:M][M:L] = [G:M]|N|$ is a π'-number, so $L \in \mathrm{Hall}_\pi(G)$. Note that K and N are complements in KN, since $p \notin \pi$. Also $|L \cap KN| = |L||KN|/|LKN| = |L||K||N|/|M| = |K|$, so $L \cap KN$ is another complement to N in KN, which is solvable. Hence, by Theorem 4.2.5, $\exists x \in KN$ with $K = (L \cap KN)^x$. Thus $K \leq L^x \in \mathrm{Hall}_\pi(G)$.

The conjugacy conclusion remains to be proved. Let us again use induction on $|G|$. Suppose that H, $K \in \mathrm{Hall}_\pi(G)$, and choose N minimal normal in G, a p-group as above. Then HN/N and KN/N are in $\mathrm{Hall}_\pi(G/N)$ so they are conjugate by induction and $\exists x \in G$ with $H^x \leq KN$. If $p \in \pi$ then KN is a π-group, so $KN = K$ (since $K \in \mathrm{Hall}_\pi(G)$), and $H^x = K$ in that case. Suppose, finally, that $p \notin \pi$. Then N is a normal Hall subgroup of KN, and it has K and H^x as complements. Apply Theorem 4.2.5 again to see that K and H^x are conjugate, and hence so are K and H. $\triangle$

The first group for which the conclusions of Theorem 4.2.6 could fail is $\mathrm{Alt}(5)$, and indeed $\mathrm{Alt}(5)$ has neither a Hall $\{2,5\}$-subgroup nor a Hall $\{3,5\}$-subgroup. (Why?) That is no accident, as in fact Hall also proved a rather strong converse to the theorem, providing a characterization of finite solvable groups.

The proof (below) requires another borrowing of a theorem of Burnside (the $p^a q^b$ Theorem, 5.2.39) from a later chapter. This practice is perhaps less than satisfying from a pedagogical point of view, but it presents no logical difficulties, as Hall's Theorem will not be used in the proof of Burnside's Theorem.

Hall's converse to Theorem 4.2.6 does not require the existence of Hall π-subgroups for all sets π of primes, only of p-complements (not necessarily normal) for all $p \in \pi(G)$, which are of course just Hall $\{p\}'$-subgroups.

Exercise

Suppose that $p \neq q$ in $\pi(G)$, and suppose that H and K are a p-complement and a q-complement, respectively, in G. Show that $H \cap K$ is a p-complement in K and a q-complement in H, and show that $H \cap K \in \mathrm{Hall}_{\{p,q\}'}(G)$. (*Hint* See Exercise I.2.2 of [31].)

Theorem 4.2.7 (P. Hall) *If G has a p-complement for every p in $\pi(G)$ then G is solvable.*

Proof Suppose that the theorem is false, and choose a nonsolvable group G of minimal order having p-complements for all p. The Burnside $p^a q^b$ Theorem (5.2.39) asserts that groups whose orders have only 2 prime divisors are solvable, so $\pi(G) \geq 3$. Fix $p \in \pi(G)$, and let H be a p-complement in G. Then H has a q-complement for all $q \in \pi(H) = \pi(G)\backslash\{p\}$ by the exercise preceding the theorem, so H is solvable, since $|G|$ is minimal. Choose a minimal normal subgroup N of H, an r-group for some prime r by Proposition 4.2.4. Choose $q \neq p, r$ in $\pi(G)$ and let K be a q-complement in G; K is solvable just as H is. If $R \in \mathrm{Syl}_r(K)$ then $\exists x \in G$ so that $N \leq R^x \leq K^x$, also a q-complement. Thus we may replace K by K^x and assume that $N \leq K$.

Set $L = \langle \cup\{N^x : x \in G\}\rangle \triangleleft G$. Since $G = HK$ we have

$$L = \langle \cup\{N^{hk} : h \in H, \ k \in K\}\rangle = \langle \cup\{N^k : k \in K\}\rangle \leq K,$$

so L is solvable. But now G/L satisfies the hypotheses of the theorem, so it is solvable (since $|G|$ is minimal), forcing G to be solvable, a contradiction. $\triangle$

4.3 Mostly p-groups

If G is any group then its *Frattini subgroup* $\Phi(G)$ is defined to be the intersection of all maximal (proper) subgroups of G (if G has no maximal subgroups then $\Phi(G) = G$). If G is finite (which will be assumed throughout this section), and $G \neq 1$, then clearly $\Phi(G) < G$. Furthermore, since any $\varphi \in \mathrm{Aut}\,G$ permutes the maximal subgroups of G, it is clear that $\Phi(G)\,\mathrm{char}\,G$.

Proposition 4.3.1 *Suppose that $H \triangleleft G$. Then $G = HK$ for some $K < G$ if and only if $H \not\leq \Phi(G)$.*

Proof We have $H \not\leq \Phi(G)$ if and only if $H \not\leq K$ for some maximal $K < G$, in which case $G = HK$. $\triangle$

A subgroup $K < G$ with $G = HK$, as in the proposition, is sometimes called a *partial complement* for H in G.

An element y in a group G is called a *nongenerator* if for any subset $X \subseteq G$ with $G = \langle X \cup \{y\}\rangle$ it is also true that $G = \langle X \rangle$. For example $y = 1$ is a nongenerator in any group; if $G = \mathrm{Sym}(3)$ then 1 is the only nongenerator. In the quaternion group of order 8 the unique element of order 2 is a nongenerator.

Proposition 4.3.2 *The Frattini subgroup $\Phi(G)$ is the set of nongenerators in G.*

Proof Take $y \in \Phi(G)$, so $y \in M$ for all maximal subgroups M. If $G \neq \langle X \rangle$ but $G = \langle X \cup \{y\} \rangle$ choose a maximal subgroup $M < G$ with $M \geq \langle X \rangle$. Then $y \in M$, so $G \leq M$, a contradiction. Conversely, suppose that y is a nongenerator and that $M < G$ is maximal. If $y \notin M$ then $\langle M \cup \{y\} \rangle = G$ but $\langle M \rangle \neq G$, also a contradiction. $\triangle$

Exercise

If $G/\Phi(G)$ is cyclic show that G is cyclic. What is $\Phi(G)$ if G is cyclic?

Proposition 4.3.3 *The Frattini subgroup $\Phi(G)$ is nilpotent.*

Proof It is sufficient to show that all Sylow subgroups of $\Phi(G)$ are normal. Take $P \in \mathrm{Syl}_p(\Phi(G))$. Then $G = N_G(P)\Phi(G)$ by the Frattini Argument (Proposition 4.2.1), so $G = N_G(P)$ by Proposition 4.3.1, i.e. $P \triangleleft G$ and so $P \triangleleft \Phi(G)$. $\triangle$

Proposition 4.3.4 *Suppose that G is a p-group for some prime p. Then*

1. $G' \leq \Phi(G)$,

2. $G/\Phi(G)$ is elementary abelian, and

3. $\Phi(G) = 1$ if and only if G is elementary abelian.

Proof Each maximal $M < G$ is normal and of index p, so G/M is abelian and $G' \leq M$; hence $G' \leq \Phi(G)$. Furthermore $x^p \in M$ for all $x \in G$ and all maximal subgroups M, so $x^p \in \Phi(G)$. Now $G/\Phi(G)$ is abelian and the pth power of each of its elements is the identity, so it is elementary abelian. It is immediate then that G is elementary abelian if $\Phi(G) = 1$. On the other hand, if G is elementary abelian it is clear that its maximal subgroups intersect in 1. $\triangle$

Corollary 4.3.5 *If G is a p-group then $\Phi(G/\Phi(G))$ is trivial.*

Exercise

If G is a p-group, $N \triangleleft G$, and G/N is elementary abelian show that $N \geq \Phi(G)$; conclude that $\Phi(G)$ is the smallest normal subgroup yielding an elementary abelian quotient.

If G is a p-group then by Proposition 4.3.4 it is clear that $V = G/\Phi(G)$ can be viewed as a vector space over the field $F = \mathbb{F}_p$ with p elements. That is the point of view of the next theorem.

Theorem 4.3.6 (Burnside's Basis Theorem) *Suppose that G is a finite p-group, write Φ for $\Phi(G)$, and set $V = G/\Phi$, viewed as an $F = \mathbb{F}_p$ vector space. Suppose that $[G:\Phi] = p^n$. Then n is the minimal size for sets of generators for G; if $X = \{x_1, \ldots, x_n\} \subseteq G$ and $v_i = \Phi x_i \in V$, $1 \le i \le n$, then $G = \langle X \rangle$ if and only if $\{v_1, \ldots, v_n\}$ is a basis for V.*

Proof Clearly $\dim_F(V) = n$. Thus V has a basis consisting of n vectors and has no smaller generating sets. The set $\{v_1, \ldots, v_n\}$ is a basis for V if and only if it generates V, which is if and only if $G = \langle X \cup \Phi(G) \rangle = \langle X \rangle$, by Proposition 4.3.2. Everything follows. $\triangle$

Corollary 4.3.7 *If G is a p-group with a unique subgroup H of index p then G is cyclic.*

Proof This is an immediate consequence of the Basis Theorem, since $\Phi(G) = H$; hence $\dim_F(G/\Phi) = 1$, and G has 1 generator. However, a simple alternative proof is also available. If G is not cyclic then every $\langle x \rangle$, $x \in G$, is contained in a maximal subgroup, of which H is the only one, so $G \le H$, a contradiction. $\triangle$

Exercises

1. Let $G = Q_m = \langle a, b \mid a^{2m} = 1, b^2 = a^m, b^{-1}ab = a^{2m-1} \rangle$, the (generalized) quaternion group of order $4m$, as on page 30. Set $A = \langle a \rangle$. Show that if $x \in G \backslash A$ then $|x| = 4$; conclude that G has a unique element of order 2.

2. Show that Q_m has the presentation $\langle a, b \mid a^m = b^2 = (ab)^2 \rangle$.

The next few results go back to Frobenius and Burnside (see Chapter VIII of [10]); the exposition below is based on that in [70], with further inspiration from [54].

We begin with some slightly technical commutator identities.

Proposition 4.3.8 *Suppose that G is a group with $G' \le Z(G)$, $x, y \in G$, and $m, n \in \mathbb{Z}$. Then*

1. $[x, y]^m = [x^m, y] = [x, y^m]$,
2. $[x, y^m][x, y^n] = [x, y^{m+n}]$, and
3. $(xy)^m = x^m y^m [x, y^{-(m-1)m/2}]$.

Proof Note that

$$[x, y][x^{-1}, y] = x^{-1}[x^{-1}, y]y^{-1}xy = x^{-1}xy^{-1}x^{-1}yy^{-1}xy = 1.$$

Thus $[x, y]^{-1} = [x^{-1}, y]$, anycd it will suffice to assume that $m > 0$ and argue by induction.

For Part 1 we have

$$[x,y]^{m+1} \;=\; [x,y][x,y]^m = [x,y][x^m,y]$$

$$=\; x^{-1}[x^m,y]y^{-1}xy = x^{-1}x^{-m}y^{-1}x^m yy^{-1}xy = [x^{m+1},y],$$

and also

$$[x,y]^m = [y,x]^{-m} = [y^m,x]^{-1} = [x,y^m].$$

For Part 2

$$[x,y^m][x,y^n] \;=\; x^{-1}y^{-m}x[x,y^n]y^m$$

$$=\; x^{-1}y^{-m}xx^{-1}y^{-n}xy^n y^m = [x,y^{m+n}].$$

Finally, for Part 3,

$$(xy)^{m+1} \;=\; (xy)^m xy = x^m y^m xy[x,y^{-(m-1)m/2}]$$

$$=\; x^{m+1}(x^{-1}y^m xy^{-m})y^{m+1}[x,y^{-(m-1)m/2}]$$

$$=\; x^{m+1}y^{m+1}[x,y^{-m}][x,y^{-(m-1)m/2}]$$

$$=\; x^{m+1}y^{m+1}[x,y^{-m(m+1)/2}],$$

by Part 2.   △

Theorem 4.3.9 *Suppose that G is a p-group having two different cyclic subgroups of index p but only one subgroup of order p. Then G is the quaternion group Q of order 8.*

Proof Note that G is not abelian. Say that $|G| = p^m$, with $[G{:}A] = [G{:}B] = p$, $A = \langle a \rangle \neq B = \langle b \rangle$. Then $A \triangleleft G$, $B \triangleleft G$, and $G = \langle a, b \rangle$. If $C = A \cap B$ then $[G{:}C] = p^2$, $C \le Z(G)$, and $C \ge G'$ since G/C is abelian. In fact, $Z(G) = C$, since $Z(G)$ cannot have index p. Also $G/C = \langle Ca, Cb \rangle$ is not cyclic, so $x^p \in C$ for all $x \in G$.

If p is odd and $x, y \in G$ then $[x, y^{p(p-1)/2}] = 1$, since $y^p \in Z(G)$. Then by Proposition 4.3.8 we have $(xy)^p = x^p y^p$. Thus the map $\varphi{:}x \mapsto x^p$ is a homomorphism; its kernel is $\{x \in G{:}x^p = 1\}$. But $[G{:}\ker \varphi] = |\operatorname{Im} \varphi| \le p^{m-2}$, so $|\ker \varphi| \ge p^2$, and $\ker \varphi$ has more than one subgroup of order p.

Thus $p = 2$. Say that $C = \langle c \rangle$; since both a^2 and b^2 generate C we may choose a, b, and c so that $a^2 = c$ and $b^2 = c^{-1}$. Thus $(ab)^4 = a^4 b^4[a, b^{-6}] = 1$, but $(ab)^2 = a^2 b^2[a, b^{-1}] = [a, b^{-1}] \neq 1$, since G is not abelian. So $|ab| = 4$, and clearly $ab \notin C$. If $|C| > 2$ choose $d \in C$ having order 4. Then $\langle ab, d \rangle$ is abelian and not cyclic, so it has 2 distinct elements of order 2. Conclude that $|C| = 2$, so $|G| = 8$, and $G \cong Q$ since D_4 has 5 elements of order 2. Alternatively, observe that $(ab)^2 = c$, so $a^2 = b^2 = (ab)^2$, and apply the exercise on page 91.   △

Exercises

1. If $N \triangleleft G$ show that $C_G(N) \triangleleft G$.

2. If G is finite and nilpotent, and $N \triangleleft G$, show that $N \cap Z(G) \neq 1$.

Proposition 4.3.10 *If G is finite and nilpotent, but not abelian, and A is a maximal abelian normal subgroup of G, then $C_G(A) = A$.*

Proof Suppose not; write $C = C_G(A)$ and note that $C \triangleleft G$. Since G/A is nilpotent there is an $x \in G$ such that $A \neq Ax \in (C/A) \cap Z(G/A)$, by the exercise above. Thus $\langle A \cup \{x\}\rangle/A \triangleleft G/A$, so $\langle A \cup \{x\}\rangle \triangleleft G$, and $\langle A \cup \{x\}\rangle$ is abelian, since $x \in C$, so the maximality of A has been contradicted. $\triangle$

Theorem 4.3.11 *Suppose that $|G| = p^m$ for some prime p and that G has a unique subgroup of order p. Then G is either cyclic or generalized quaternion.*

Proof Suppose first that p is odd and apply induction on m. The result is clear for small m. Suppose then that $m > 1$, and assume the result for lower powers. Thus every subgroup of index p is cyclic, so G is cyclic by Corollary 4.3.7 and Theorem 4.3.9.

Suppose next that $p = 2$. The result is clear if G is abelian, so suppose also that it is not abelian, and let A be a maximal abelian normal subgroup. Thus A is cyclic, and $C_G(A) = A$ by Proposition 4.3.10. Choose $b \in G \backslash A$ with $b^2 \in A$; note that $A \neq \langle b^2 \rangle$ since $b \notin C_G(A)$. Choose $C = \langle c \rangle \leq A$ with $[C: \langle b^2 \rangle] = 2$, and set $H = \langle b, c \rangle$, $B = \langle b \rangle$. Now $B \cap C = \langle b^2 \rangle$, so $|H| = |B||C|/|\langle b^2 \rangle| = 2|C|$, or $[H:C] = 2$. Also $[H: \langle b^2 \rangle] = [H:C][C: \langle b^2 \rangle] = 4$, so $[H:B] = 2$. Thus H has a unique subgroup of order 2 but 2 subgroups, B and C, of index 2, so H is quaternion of order 8 by Theorem 4.3.9; in particular $|b| = |c| = 4$.

Say that $|a| = 2^{n+1}$; then $a^{2^n} = b^2$, since both have order 2. The argument can be repeated with b replaced by ab, since $ab \notin A$ but $(ab)^2 = a^2b^2[a, b^{-1}] \in A$. Thus also $a^{2^n} = (ab)^2$, and $\langle a, b \rangle$ is a homomorphic image of Q_{2^n} by the exercise on page 91. Since $|\langle a, b \rangle| \geq 2^{n+2}$ we have $\langle a, b \rangle \cong Q_{2^n}$.

It remains to be shown that $G = \langle a, b \rangle$, and it will suffice to show that $[G:A] = 2$. If that is not the case choose $x \in G$ with $x^2 \notin A$ but $x^4 \in A$. By the argument above $\langle a, x^2 \rangle \cong Q_{2^n}$, and in particular $x^2 a x^{-2} = a^{-1}$. Let θ denote the homomorphism (Proposition 4.1.10) from G to Aut A that restricts inner automorphisms. Say that $\theta_x(a) = a^k$, and note that k is odd since $\langle a^k \rangle = A$. Thus $k^2 \equiv 1 \pmod{4}$. But also $\theta_x^2(a) = a^{k^2} = \theta_{x^2}(a) = a^{-1}$, so $k^2 \equiv -1 \pmod{2^{n+1}}$; hence $k^2 \equiv -1 \pmod{4}$, a contradiction. We conclude that $x^2 \in A$ for all $x \in G \backslash A$ and also that $\theta_x(a) = a^{-1}$. Thus the image of θ in Aut A has order 2, and ker $\theta = C_G(A) = A$, so $[G:A] = 2$. $\triangle$

Corollary 4.3.12 *Suppose that* $|G| = p^m$, $1 < k < m$, *and that* G *has a unique subgroup* H *of order* p^k. *Then* G *is cyclic.*

Proof Choose $K \leq G$ with $[K:H] = p$, then K is cyclic by Corollary 4.3.7, hence H is also cyclic. Any $L \leq G$ of order p is contained in a subgroup of order p^k, therefore in H, so G has only one subgroup of order p, and is either cyclic or generalized quaternion by the theorem. Similarly any $M \leq G$ of order p^2 is contained in a subgroup of order p^k, therefore in H, so G has only one subgroup of order p^2. But Q_{2^n} has more than one subgroup of order 4, so G must be cyclic. $\triangle$

Exercise

If G is a finite p-group show that G has exactly one subgroup of order p if and only if every subgroup of order p^2 is cyclic.

Proposition 4.3.13 *Suppose that* G *is a group,* x, $y \in G$, $x^2 = y^2 = 1$, *and that* $1 \neq H = \langle x, y \rangle$ *is finite. Then* H *is a dihedral group.*

Proof If $x = 1$ or $y = 1$ or $x = y$ then $H \cong D_1$, so suppose that $1 \neq x \neq y \neq 1$. Set $z = xy$ and say that $|z| = m$; note that $H = \langle z, x \rangle$. Since $x \neq y$ at least one or the other of x and y is not in $\langle z \rangle$, so $|H| \geq 2m$. Also $xzx = x(xy)x = yx = z^{-1}$, so z and x satisfy defining relations for D_m; hence $H \cong D_m$. $\triangle$

Corollary 4.3.14 *A nontrivial homomorphic image of a dihedral group is dihedral.*

Corollary 4.3.15 *A nontrivial homomorphic image of the generalized quaternion 2-group* Q_{2^n} *is either isomorphic with* Q_{2^n} *or is dihedral.*

Proof Take $Q_{2^n} = \langle a, b \mid a^{2^{n+1}} = 1, b^2 = a^{2^n}, b^{-1}ab = a^{-1} \rangle$, as usual. Suppose that θ is a homomorphism from Q_{2^n} onto H, and assume that $1 \neq K = \ker \theta$. Then the unique element $a^{2^n} = b^2$ of order 2 is in K, so $\theta(b)^2 = \theta(b^2) = 1$. We have $H = \langle \theta(a), \theta(b) \rangle$, with $\theta(a)^{2^n} = \theta(b)^2 = 1$ and $\theta(b)^{-1}\theta(a)\theta(b) = \theta(a)^{-1}$, so H is a homomorphic image of D_{2^n}. Apply Corollary 4.3.14. $\triangle$

Some of the results above will be applied in a later chapter in the study of Frobenius groups. For the present let us give some applications to Sylow subgroups.

If $p \in \mathbb{N}$ is a prime set $q = p^k$ and let $F = \mathbb{F}_q$. Recall that $|\mathrm{SL}(2, q)| = (q - 1)q(q + 1)$. The subgroup

$$P = \left\{ \begin{bmatrix} 1 & 0 \\ a & 1 \end{bmatrix} : a \in F \right\}$$

has order q, so P is a Sylow p-subgroup of $\mathrm{SL}(2,q)$. Clearly P is isomorphic with the additive group F, which, being a vector space over its prime field, is an elementary abelian p-group.

Suppose now that p is odd. An easy calculation shows that if $\tau \in \mathrm{SL}(2,q)$ and $\tau^2 = 1$ then in fact $\tau = \pm 1$, so $\mathrm{SL}(2,q)$ has a unique element of order 2. By Theorem 4.3.11 a Sylow 2-subgroup T of $\mathrm{SL}(2,q)$ is either cyclic or generalized quaternion.

If $q = 3$ recall that $\mathrm{PSL}(2,3) \cong \mathrm{Alt}(4)$, whose Sylow 2-subgroup is a Klein 4-group. Thus the Sylow 2-subgroup of $\mathrm{SL}(2,3)$ is (isomorphic to) $Q = Q_2$. If $q \geq 5$ then $\mathrm{PSL}(2,q)$ is simple (Theorem 2.6.13), and its Sylow 2-subgroups, which are homomorphic images of T, cannot be cyclic by Corollary 4.1.12. Thus $T \cong Q_{2^m}$ for some m in all cases.

If T is a Sylow 2-subgroup of $\mathrm{SL}(2,q)$ then $T/\{\pm 1\}$ is a Sylow 2-subgroup of $\mathrm{PSL}(2,q)$ – it is dihedral by Corollary 4.3.15.

Proposition 4.3.16 *Suppose that G has more than one Sylow 2–subgroup, and that if $T_1 \neq T_2$ in $\mathrm{Syl}_2(G)$ then $T_1 \cap T_2 = 1$. Then G has exactly one conjugacy class of involutions.*

Proof Choose involutions x and y from two different Sylow 2-subgroups. Then $H = \langle x, y \rangle \cong D_m$ for some m by Proposition 4.3.13. If m is odd then x and y are conjugate in H. Could m be even? If so then $z = (xy)^{m/2} \in Z(H)$, and $|z| = 2$. But then $H_1 = \langle x, z \rangle$ and $H_2 = \langle z, y \rangle$ are both Klein 4-groups, and each is contained in a Sylow 2-subgroup, say $H_i \leq T_i$. Then $T_1 \neq T_2$ by the initial choice of x and y, but $1 \neq z \in T_1 \cap T_2$, a contradiction. $\triangle$

Suppose now that $q = 2^k$. As noted above

$$T = \left\{ \begin{bmatrix} 1 & 0 \\ a & 1 \end{bmatrix} : a \in \mathbb{F}_q \right\}$$

is a Sylow 2-subgroup of $G = \mathrm{SL}(2,q)$, and it is elementary abelian of order q. An easy calculation shows that the normalizer of T is

$$N = \left\{ \begin{bmatrix} a & 0 \\ b & a^{-1} \end{bmatrix} : a \in F^\star, \ b \in F \right\},$$

which has order $q(q-1)$. Thus there are $[G \colon N] = q + 1$ different Sylow 2-subgroups in G. The elements of order 2 are all matrices $\begin{bmatrix} 1 & 0 \\ a & 1 \end{bmatrix}$, $a \in F^\star$, of which there are $q - 1$, and all $\begin{bmatrix} a & b \\ c & a \end{bmatrix}$, where $a \in F$ and $b \in F^\star$ are arbitrary, but $c = (1 + a^2)/b$ (verify this), so there are $q(q-1)$ such matrices. All told, then, there are $q^2 - 1 = (q-1)(q+1)$ involutions in G. Since they are distributed among the $q + 1$ Sylow 2-subgroups, $q - 1$ in each of them, we see that if $T_1 \neq T_2$ in $\mathrm{Syl}_2(G)$ then $T_1 \cap T_2 = 1$. By Proposition 4.3.16 we conclude that all involutions in $\mathrm{SL}(2, 2^k)$ are conjugate.

Chapter 5

Representations and Characters

In this chapter we introduce the basic ideas of the (ordinary) representations and characters of finite groups, concentrating on characters. The applications are legion.

Group theoretical applications of character theory appear prominently in this chapter as well as later in the book, especially in Chapter 9. Some of the many other examples of applications of character theory appear in the books by Sternberg [59] (physics), James and Liebeck [41] (chemistry), and Diaconis [20] (probability and statistics).

5.1 Representations

If F is a field and V is a finite dimensional vector space over F write $\mathrm{GL}(V)$ for the *general linear group* of all invertible linear transformations of V. If $\dim_F(V) = n$, and a basis is chosen for V then we obtain an isomorphism (via representing matrices) from $\mathrm{GL}(V)$ to $\mathrm{GL}(n, F)$, the group of $n \times n$ invertible matrices over F.

An *F-representation* of a group G is a homomorphism T from G to $\mathrm{GL}(V)$ for some V. If a basis is chosen then the composite map

$$G \xrightarrow{T} \mathrm{GL}(V) \to \mathrm{GL}(n, F)$$

is called an *F-matrix representation*; it will be denoted by $\widehat{T}$.

If T is a representation of G and T is 1–1 then it is called *faithful*. As a more-or-less general rule, any adjective that applies to a representation T applies as well to any corresponding matrix representation $\widehat{T}$.

If S and T are F-representations of G on spaces V and W then we say that S and T are *equivalent*, and write $S \sim T$, if there is an F-isomorphism

$\theta \colon V \to W$ for which the diagram

$$
\begin{array}{ccc}
V & \xrightarrow{S(x)} & V \\
\theta \downarrow & & \downarrow \theta \\
W & \xrightarrow{T(x)} & W
\end{array}
$$

is commutative, i.e. $\theta S(x) = T(x)\theta$, or $T(x) = \theta S(x)\theta^{-1}$, for all $x \in G$.

If T is a representation of G on V, then a subspace W of V is called *T-invariant* if $T(x)W \subseteq W$ for all $x \in G$. Note that then each $T(x)$ can be restricted to a linear transformation of W, resulting in a representation of G on W. Denote it by $T|_W$.

If W is T-invariant, and if a basis is chosen for W and extended to a basis for V then the resulting matrix representation $\widehat{T}$ has the form

$$
\widehat{T}(x) = \left[\begin{array}{cc} A(x) & C(x) \\ 0 & B(x) \end{array} \right],
$$

where $A = \widehat{T|_W}$.

Remark B is also a matrix representation of G, corresponding to the action of G (via T) on the quotient space V/W.

If V has a proper nonzero T-invariant subspace W then T is called *reducible*, otherwise T is *irreducible*, or *simple*.

Warning! Just as for polynomials, the question of reducibility versus irreducibility of representations depends (heavily) on the field F of scalars.

Exercise

Take $G = C_4 = \langle x \mid x^4 = 1 \rangle$, cyclic of order 4, and define a (matrix) representation via $\widehat{T}(x) = \left[\begin{array}{cc} 0 & -1 \\ 1 & 0 \end{array} \right]$. If $F = \mathbb{R}$, the reals, show that T is irreducible. If $F = \mathbb{C}$, the complexes, show that T is reducible.

Suppose that S and T are F-representations of G on V and W, respectively. Define $S \oplus T$ on $V \oplus W$ via $(S \oplus T)(x) \colon (v, w) \mapsto (S(x)v, T(x)w)$ for all $x \in G$, $v \in V$, $w \in W$. Then $S \oplus T$ is an F-representation of G, called the *direct sum* of S and T. In like manner we may of course define the direct sum $T_1 \oplus T_2 \oplus \cdots \oplus T_k$ of any finite number of F-representations.

If T is an F-representation of G and T is equivalent with a direct sum of some number of irreducible F-representations then T is called *completely reducible* (or *semisimple*). Note that as a special case an irreducible representation is completely reducible.

If T is an F-representation of G on V and $\dim_F(V) = n$ we say that T has *degree* n, and write $\deg(T) = n$. If $\deg(T) = 1$ then T is called a *linear* representation. If T is linear then $\widehat{T}$ is a homomorphism from G to the multiplicative group $F^* = F \setminus \{0\}$.

Examples

1. For any G, any F, and any V define $T(x) = 1_V$, the identity transformation on V, hence $\widehat{T} = I$, an identity matrix, for all $x \in G$. Then T is called a *trivial* representation. If T is trivial and $\deg(T) = 1$ then T is called the *principal* F-representation of G and is usually denoted 1_G.

2. Let $G = C_n = \langle x \mid x^n = 1 \rangle$, take $F = \mathbb{C}$, and let $\zeta \in \mathbb{C}$ be a primitive nth root of unity, e.g. $\zeta = e^{2\pi i/n}$. Define $\widehat{T}(x) = \zeta$. Then T is a faithful irreducible linear representation of G.

3. Let $G = S_n$, the symmetric group of degree n, or more generally any subgroup of S_n, and let F be any field. Define

$$\widehat{T}(x) = \begin{cases} 1 & \text{if } x \text{ is even,} \\ -1 & \text{if } x \text{ is odd.} \end{cases}$$

Then T is called the *alternating* representation of G.

4. Suppose that G acts as a permutation group on a set $X = \{x_1, \ldots, x_n\}$. For any field F let V be a vector space over F of dimension n, and let $\{v_1, \ldots, v_n\}$ be a basis. If $g \in G$ and $x_j^g = x_i$ define $T(g)v_i = v_j$, determining $T(g) \in \mathrm{GL}(V)$ and hence an F-representation T. It is called the *permutation* representation corresponding to the action of G on X. Note that each $\widehat{T}(g)$ is a *permutation matrix*, i.e. each row and each column has exactly one nonzero entry, which is a 1.

5. In 4 above take $X = G$, and let G act on X by right multiplication (as in Cayley's Theorem). The resulting permutation representation of G is called the *right regular representation* of G; denote it by R. Similarly the permutation action of G on X by *left* multiplication results in a permutation representation called the *left regular representation*, denoted L.

Exercise

Write out the right regular (matrix) representation for generators for the groups C_4, $C_2 \times C_2$, S_3, and D_4 (dihedral of order 8).

Theorem 5.1.1 (Maschke's Theorem) *Suppose that G is a finite group, F is a field, and $\mathrm{char}(F) \nmid |G|$. Then every F-representation T of G is completely reducible.*

Proof We use induction on $n = \deg(T)$. If $n = 1$ then T is irreducible, hence completely reducible. Assume then that $n > 1$, and assume the result for all F-representations of degree less than n. Assume that T is reducible and choose a basis so that

$$\widehat{T}(x) = \begin{bmatrix} A(x) & C(x) \\ 0 & B(x) \end{bmatrix},$$

all $x \in G$. If $x, y \in G$ then, since $\widehat{T}$ is a representation, we have $\widehat{T}(xy) = \widehat{T}(x)\widehat{T}(y)$, which says that

$$\begin{bmatrix} A(xy) & C(xy) \\ 0 & B(xy) \end{bmatrix} = \begin{bmatrix} A(x)A(y) & A(x)C(y) + C(x)B(y) \\ 0 & B(x)B(y) \end{bmatrix}.$$

Thus $C(xy) = A(x)C(y) + C(x)B(y)$, and

$$C(xy)B((xy)^{-1})B(x) = C(xy)B(y^{-1}) = A(x)C(y)B(y^{-1}) + C(x). \qquad (*)$$

Define $D = |G|^{-1}\Sigma\{C(z)B(z^{-1}): z \in G\}$, and sum the equation $(*)$ above over all y in G to obtain

$$|G|DB(x) = |G|A(x)D + |G|C(x).$$

Hence

$$DB(x) = A(x)D + C(x), \quad \text{all } x \in G,$$

since $|G| \neq 0$ in F. Observe now that

$$\begin{bmatrix} I & D \\ 0 & I \end{bmatrix}\begin{bmatrix} A(x) & 0 \\ 0 & B(x) \end{bmatrix} = \begin{bmatrix} A(x) & DB(x) \\ 0 & B(x) \end{bmatrix}$$
$$= \begin{bmatrix} A(x) & A(x)D + C(x) \\ 0 & B(x) \end{bmatrix}$$
$$= \begin{bmatrix} A(x) & C(x) \\ 0 & B(x) \end{bmatrix}\begin{bmatrix} I & D \\ 0 & I \end{bmatrix}.$$

It follows that $\widehat{T} \sim A \oplus B$, and the induction hypothesis can be applied to A and B to complete the proof. $\triangle$

Proposition 5.1.2 (Schur's Lemma) *Suppose that S and T are irreducible F-representations of G on V and W, respectively, and that $A: V \to W$ is a linear transformation such that $AS(x) = T(x)A$ for all $x \in G$. Then either $A = 0$ or else A is an isomorphism (and hence $S \sim T$).*

Proof Suppose that $A \neq 0$. Set $V_1 = \ker(A)$ and $W_1 = Im(A)$. If $v \in V_1$ then $0 = T(x)Av = AS(x)v$, all x, so $S(x)v \in \ker(A)$ and V_1 is S-invariant. But $V_1 \neq V$, so $V_1 = 0$ since S is irreducible, and hence A is 1–1.

If $w \in W_1$ write $w = Au$, $u \in V$. Then $T(x)w = T(x)Au = A(S(x)u) \subseteq Im(A) = W_1$, all x. Thus W_1 is T-invariant and nonzero, so $W_1 = W$ since T is irreducible, and A is onto. △

If T is an F-representation of G on V define the *centralizer* of T to be the algebra of all linear transformations $A: V \to V$ for which $AT(x) = T(x)A$ for all $x \in G$. We write $\mathcal{C}(T)$ or $\mathcal{C}_F(T)$ for the centralizer of T. The centralizer $\mathcal{C}(\widehat{T})$ of a corresponding matrix representation $\widehat{T}$ is the algebra of all matrices over F that commute with all $\widehat{T}(x)$, $x \in G$.

Proposition 5.1.3 (1) *If T is an irreducible F-representation of G on V then $\mathcal{C}(T)$ is a division algebra.* (2) *If* $\mathrm{char}(F) \nmid |G|$ *and $\mathcal{C}(T)$ is a division algebra then T is irreducible.*

Proof (1) If $A \neq 0$ in $\mathcal{C}(T)$ then A is an automorphism of V by Schur's Lemma (5.1.2), so A^{-1} exists. If $x \in G$ then $A^{-1}T(x) = (T(x^{-1})A)^{-1} = (AT(x^{-1}))^{-1} = T(x)A^{-1}$, so $A^{-1} \in \mathcal{C}(T)$ and $\mathcal{C}(T)$ is a division algebra.

(2) Suppose that T is reducible. By Maschke's Theorem (5.1.1) $V = V_1 \oplus V_2$, with both V_i nonzero and T-invariant, so we may assume that $T = T_1 \oplus T_2$, with $T_i = T|_{V_i}$. Let P be the projection of V onto V_1 along V_2, i.e. if $v = v_1 + v_2 \in V$, with $v_i \in V_i$, then $Pv = v_1$. Note that P is *not* invertible, since $\ker(P) = V_2 \neq 0$. However,

$$PT(x)v = PT(x)(v_1 + v_2) = P(T(x)v_1 + T(x)v_2) = T(x)v_1 = T(x)Pv,$$

so $P \in \mathcal{C}(T)$, contradicting the assumption that $\mathcal{C}(T)$ is a division algebra. △

Exercise

Let $G = C_3 = \langle x \mid x^3 = 1 \rangle$ and set $\widehat{T}(x) = \begin{bmatrix} 0 & -1 \\ 1 & -1 \end{bmatrix}$. Compute $\mathcal{C}(T)$. If $F \subseteq \mathbb{R}$ show that T is irreducible. If $F = \mathbb{C}$ find a matrix M such that $M^{-1}\widehat{T}(x)M = \begin{bmatrix} \omega & 0 \\ 0 & \omega^2 \end{bmatrix}$, where $\omega^2 + \omega + 1 = 0$ in $\mathbb{C}$.

If V is a vector space over F and K is an extension field of F we may set $V^K = K \otimes_F V$ to extend the field of scalars from F to K. Recall that if $\{v_1, \ldots, v_n\}$ is an F-basis for V then $\{1 \otimes v_1, \ldots, 1 \otimes v_n\}$ is a K-basis for V^K. If T is an F-representation of G on V define T^K on V^K via $T^K(x): 1 \otimes v_i \mapsto 1 \otimes T(x)v_i$, $1 \leq i \leq n$, all $x \in G$. Note that, relative to the bases indicated, $\widehat{T^K} = \widehat{T}$. This process is referred to as *extending the ground field*.

If T is an irreducible F-representation of G and if T^K remains irreducible for every extension field K of F, then T is said to be *absolutely irreducible*. If F is a field for which every irreducible F-representation is absolutely irreducible then F is called a *splitting field* for G.

Proposition 5.1.4 (1) *If T is an absolutely irreducible F-representation of G then $\mathcal{C}(T) = \{\lambda 1 \colon \lambda \in F\} \cong F$. (2) If $\operatorname{char}(F) \nmid |G|$, and if $\mathcal{C}(T) = \{\lambda 1 \colon \lambda \in F\}$, then T is absolutely irreducible.*

Proof (1) Let $\overline{F}$ be an algebraic closure of F. Then $T^{\overline{F}}$ is irreducible and $\mathcal{C}_F(\widehat{T}) \subseteq \mathcal{C}_{\overline{F}}(\widehat{T})$. Take $A \in \mathcal{C}_F(\widehat{T})$ and let $\lambda \in \overline{F}$ be an eigenvalue of A. Thus $A - \lambda I$ is singular. Clearly $A - \lambda I \in \mathcal{C}_{\overline{F}}(\widehat{T})$, which is a division algebra by Proposition 5.1.3. Thus $A - \lambda I = 0$, so $A = \lambda I$, and in particular $\lambda \in F$.

(2) We show first that $\mathcal{C}_K(T^K) = \{\lambda 1 \colon \lambda \in K\}$ for every extension field $K \supseteq F$. Take $A \in \mathcal{C}_K(T^K)$ and choose a basis (for V^K) so that A is represented by a matrix M. Write

$$M = \alpha_1 M_1 + \cdots + \alpha_k M_k,$$

where each M_i is an F-matrix, each $\alpha_i \in K$, and the set $\{\alpha_1, \ldots, \alpha_k\}$ is linearly independent over F (why is this possible?). Then $M\widehat{T}(x) = \widehat{T}(x)M$, all $x \in G$, or $\sum_{i=1}^k \alpha_i M_i \widehat{T}(x) = \sum_{i=1}^k \alpha_i \widehat{T}(x) M_i$. Read that equality entry by entry to conclude (since $\{\alpha_i\}$ is linearly independent) that $M_i \widehat{T}(x) = \widehat{T}(x) M_i$, $1 \leq i \leq k$. Thus $M_i \in \mathcal{C}_F(\widehat{T})$, so $M_i = \lambda_i I$ for some $\lambda_i \in F$, and so $M = \sum_i \alpha_i M_i = (\sum_i \alpha_i \lambda_i) I \in \{\lambda I \colon \lambda \in K\}$. In particular we see that $\mathcal{C}_K(T^K)$ is a division algebra for all $K \supseteq F$, and so T^K is irreducible by Proposition 5.1.3. Thus T is absolutely irreducible. △

Although we will not need the more general statement it should be noted that the conclusion of part 2 of Proposition 5.1.4 holds without the assumption that $\operatorname{char}(F) \nmid |G|$. For a proof see Isaacs [39], page 145.

Corollary 5.1.5 *(Of the proof of Proposition 5.1.4.) If $\operatorname{char}(F) \nmid |G|$ and if F is algebraically closed, then F is a splitting field for G.*

Exercise

Suppose that G has a faithful irreducible $\mathbb{C}$-representation T. Show that the center $Z = Z(G)$ is cyclic. (*Hint* If $z \in Z$ then $T(z) \in \mathcal{C}(T)$.)

Proposition 5.1.6 *Suppose that S and T are F-representations of G, F is an infinite field, and K is an extension field of F. If $S^K \sim T^K$ then $S \sim T$, i.e. inequivalent representations cannot be made equivalent by extending the ground field.*

Proof Since S^K and T^K are equivalent there is an invertible K-matrix M such that $M\widehat{S}(x) = \widehat{T}(x)M$, all $x \in G$. As in the proof of Proposition 5.1.4 write $M = \sum_{i=1}^k \alpha_i M_i$, where the M_i are F-matrices and $\{\alpha_1, \ldots, \alpha_k\} \subseteq K$

is F-linearly independent. Since $M\widehat{S}(x) = \widehat{T}(x)M$ we have, as above, that $M_i\widehat{S}(x) = \widehat{T}(x)M_i$, $1 \leq i \leq k$ and all $x \in G$. Define a polynomial

$$P(t_1, \ldots, t_k) = \det(t_1 M_1 + \cdots + t_k M_k) \in F[t_1, \ldots, t_k].$$

Note that $P(\alpha_1, \ldots, \alpha_k) = \det(M) \neq 0$, so P is not the zero polynomial. Since F is infinite there are $\beta_1, \ldots, \beta_k \in F$ so that

$$P(\beta_1, \ldots, \beta_k) = \det(\beta_1 M_1 + \cdots + \beta_k M_k) \neq 0.$$

(Why?) Set $N = \sum_i \beta_i M_i$, an invertible F-matrix, and observe that $N\widehat{S}(x) = \widehat{T}(x)N$, all $x \in G$, so $S \sim T$. $\triangle$

Remark The result above is also true for finite fields F, but the proof is more complicated (e.g. see Theorem 29.7 of [18]).

Proposition 5.1.7 *Suppose that G is abelian and that T is an absolutely irreducible F-representation of G on V. Then $\deg(T) = 1$.*

Proof Since G is abelian $T(x) \in \mathcal{C}(T)$ for all $x \in G$. Thus $T(x) = \lambda_x 1$ for some $\lambda_x \in F$, all $x \in G$. But then *every* subspace of V must be T-invariant and so $\deg(T) = \dim(V) = 1$ by irreducibility. $\triangle$

Theorem 5.1.8 (Schur) (1) *Suppose that S and T are inequivalent irreducible F-representations. Choose bases so that $\widehat{S}(x) = [s_{ij}(x)]$ and $\widehat{T}(x) = [t_{ij}(x)]$, all $x \in G$. Then*

$$\sum \{s_{ij}(x)t_{kl}(x^{-1}) : x \in G\} = 0$$

for all $i, j, k,$ and l.
(2) *If T is absolutely irreducible, and $n = \deg(T)$, then*

$$n \sum \{t_{ij}(x)t_{kl}(x^{-1}) : x \in G\} = \delta_{il}\delta_{jk}|G|,$$

for all $i, j, k,$ and l.

Proof (1) For any F-matrix M (of appropriate size) define

$$L = \sum \{\widehat{S}(x)M\widehat{T}(x^{-1}) : x \in G\}.$$

If $y \in G$ then

$$\begin{aligned}
\widehat{S}(y)L &= \sum \{\widehat{S}(y)\widehat{S}(x)M\widehat{T}((yx)^{-1})\widehat{T}(y) : x \in G\} \\
&= \left[\sum_x \widehat{S}(yx)M\widehat{T}((yx)^{-1})\right]\widehat{T}(y) = L\widehat{T}(y).
\end{aligned}$$

By Schur's Lemma (5.1.2) $L = 0$ for every choice of M. Choose $M = E_{jk}$, the matrix with 1 in position jk and 0's elsewhere. Then the il-entry of L is $0 = \sum_x s_{ij}(x)t_{kl}(x^{-1})$.

(2) Set $L = \sum\{\widehat{T}(x)M\widehat{T}(x^{-1}) : x \in G\}$ for any M and check, as above, that if $y \in G$ then $\widehat{T}(y)L = L\widehat{T}(y)$, so $L \in \mathcal{C}(\widehat{T})$. Again take $M = E_{jk}$. By Proposition 5.1.4 $L = \lambda_{jk}I$ for some $\lambda_{jk} \in F$. Thus the il-entry of L is

$$
\begin{aligned}
\lambda_{jk}\delta_{il} &= \sum_x t_{ij}(x)t_{kl}(x^{-1}) = \sum_x t_{ij}(x^{-1})t_{kl}(x) \\
&= \sum_x t_{kl}(x)t_{ij}(x^{-1}) = \lambda_{li}\delta_{kj}.
\end{aligned}
$$

Choose $i = l$ and $j \neq k$ to see that $\lambda_{jk} = 0$ when $j \neq k$. Next choose $i = l$ and $j = k$ to see that $\lambda_{ii} = \lambda_{jj} = \lambda$ (say). So, $\lambda = \sum_x t_{ij}(x)t_{ji}(x^{-1})$ for all i and j. Now sum over j and get

$$
n\lambda = \sum_x \sum_j t_{ij}(x)t_{ji}(x^{-1}) = \sum_x 1 = |G|,
$$

since $\sum_j t_{ij}(x)t_{ji}(x^{-1})$ is the ii-entry of $\widehat{T}(x)\widehat{T}(x^{-1}) = \widehat{T}(1) = I_n$. Thus

$$
n\sum_{x \in G} t_{ij}(x)t_{kl}(x^{-1}) = n\lambda_{jk}\delta_{il} = (n\lambda)\delta_{jk}\delta_{il} = |G|\delta_{jk}\delta_{il}.
$$

$\triangle$

Theorem 5.1.9 (Frobenius and Schur) *Suppose that* $\mathrm{char}(F) \nmid |G|$, *and let* $T_1, \ldots, T_k$ *be mutually inequivalent absolutely irreducible F-representations of G. Choose bases so that* $\widehat{T_s}(x) = [t_{ij}^{(s)}(x)]$ *for each s and all x. Then the set* $\{t_{ij}^{(s)} : 1 \leq s \leq k,$ *all* $i, j\}$ *is an F-linearly independent set of functions from G to F.*

Proof Suppose that $\sum_{s,i,j} \alpha_{ij}^{(s)} t_{ij}^{(s)} = 0$, $\alpha_{ij}^{(s)} \in F$. Say that $\deg(T_s) = n_s$. Choose arbitrary r, l, and m, and note that

$$
\begin{aligned}
0 &= \sum_{x \in G} n_r \sum_{s,i,j} \alpha_{ij}^{(s)} t_{ij}^{(s)}(x)t_{lm}^{(r)}(x^{-1}) \\
&= \sum_{s,i,j} \alpha_{ij}^{(s)} n_r \sum_x t_{ij}^{(s)}(x)t_{lm}^{(r)}(x^{-1}) \\
&= \sum_{s,i,j} \alpha_{ij}^{(s)} |G|\delta_{rs}\delta_{jl}\delta_{im} = |G|\alpha_{ml}^{(r)},
\end{aligned}
$$

so $\alpha_{ml}^{(r)} = 0$ for all r, l, and m.

$\triangle$

Corollary 5.1.10 *If $n_s = \deg(T_s)$, all s, then $\sum_{s=1}^{k} n_s^2 \leq |G|$.*

Proof There are $n_1^2 + \cdots + n_k^2$ matrix entry functions, and the dimension of the space F^G of *all* functions from G to F is $|G|$.   △

Corollary 5.1.11 *There are at most $|G|$ inequivalent absolutely irreducible F-representations of G.*

Recall (Corollary 5.1.5) that algebraically closed fields are splitting fields. It is usually not necessary to extend all the way to an algebraic closure, however, in order to obtain a splitting field for a group G.

Proposition 5.1.12 *If $\mathrm{char}(F) \nmid |G|$ then there is a finite extension K of F that is a splitting field for G.*

Proof Let $\overline{F}$ be an algebraic closure for F, a splitting field for G by Corollary 5.1.5. Let $\{T_1, \ldots, T_k\}$ be a full set of inequivalent irreducible (hence absolutely irreducible) $\overline{F}$-representations of G. Choose bases to obtain corresponding matrix representations $\widehat{T_s}$, say with $\widehat{T_s}(x) = [t_{ij}^{(s)}(x)]$, all $x \in G$. The matrix entries $t_{ij}^{(s)}(x)$ are all algebraic over F (they are in $\overline{F}$), and there are only finitely many of them. Adjoin them all to F to obtain a field K which is a finite extension of F, and observe that K is a splitting field for G since all $\widehat{T_s}$, $1 \leq s \leq k$, are K-representations.   △

Corollary 5.1.13 *For any G there is an algebraic number field (i.e., a finite extension of $\mathbb{Q}$) that is a splitting field for G.*

5.2   Characters

Recall that the *trace* of an $n \times n$ matrix $A = [a_{ij}]$ is $\mathrm{tr}(A) = \sum_{i=1}^{n} a_{ii}$. It is easy to check that $\mathrm{tr}(AB) = \mathrm{tr}(BA)$ for all A, B In particular if M is invertible then $\mathrm{tr}(M^{-1}AM) = \mathrm{tr}(A)$. Consequently, if $T: V \to V$ is a linear transformation, we may unambiguously define $\mathrm{tr}(T) = \mathrm{tr}(A)$ for *any* representing matrix A of T.

If T is an F-representation of G define the *character* $\chi = \chi_T$ of T to be the function from G to F determined by $\chi(x) = \mathrm{tr}(T(x))$, all $x \in G$. The character of an F-representation will often be called an *F-character*. *Remark* If K is an extension field of F then the character of T^K is the same as the character of T, since $\widehat{T^K} = \widehat{T}$ for appropriate bases.

As a general rule any adjective that can be applied to a representation will be applied as well to its character. Thus χ can be faithful, reducible, irreducible, linear, etc.

Note that $\chi_T(1) = \mathrm{tr}(I) = \deg(T) \cdot 1 \in F$. Thus if $\mathrm{char}(F) = 0$ then $\chi(1)$ (as an integer) is the degree of T, and in that case we call it the *degree* of χ and write $\deg(\chi) = \chi(1)$.

Proposition 5.2.1 *If $\chi = \chi_T$ is a character of G then χ is a class function; i.e. χ is constant on conjugacy classes of G.*

Proof If $x, y \in G$ then

$$
\begin{aligned}
\chi(y^{-1}xy) &= \operatorname{tr} T(y^{-1}xy) \\
&= \operatorname{tr}[T(y^{-1})T(x)T(y)] = \operatorname{tr} T(x) = \chi(x).
\end{aligned}
$$

$\triangle$

Proposition 5.2.2 *If S and T are F-representations of G, and $S \sim T$, then $\chi_S = \chi_T$.*

Proof For appropriately chosen bases we have $\widehat{S} = \widehat{T}$. $\triangle$

Proposition 5.2.3 *If S and T are F-representations of G then*

$$\chi_{S \oplus T} = \chi_S + \chi_T.$$

Proof This is clear, since $\widehat{S \oplus T} \sim \begin{bmatrix} \widehat{S} & 0 \\ 0 & \widehat{T} \end{bmatrix}$. $\triangle$

Corollary 5.2.4 *Suppose that $\operatorname{char}(F) \nmid |G|$ and that χ is an F-character of G. Then there are irreducible F-characters $\chi_1, \ldots, \chi_k$ of G so that $\chi = \chi_1 + \cdots + \chi_k$.*

Proof Say that $\chi = \chi_T$. By Maschke's Theorem (5.1.1) there are irreducible F-representations $T_1, \ldots, T_k$ so that $T \sim T_1 \oplus \cdots \oplus T_k$. Let χ_i be the character of T_i for each i and apply Proposition 5.2.3. $\triangle$

Proposition 5.2.5 *Suppose that $F \subseteq \mathbb{C}$ and that χ is an F-character of G. Then $\chi(x^{-1}) = \overline{\chi(x)}$ for all $x \in G$.*

Proof Say that $\chi = \chi_T$. Set $H = \langle x \rangle \le G$ and restrict T to H, giving a representation of H. Extend the field to $\mathbb{C}$ (χ is unchanged). By Maschke's Theorem (5.1.1), Corollary 5.1.7, and Proposition 5.1.7 (applied to $T|_H$) we may choose a basis so that

$$
\widehat{T}(x) = \begin{bmatrix} \zeta_1 & & 0 \\ & \ddots & \\ 0 & & \zeta_k \end{bmatrix},
$$

with each $\zeta_i \in \mathbb{C}$. Say that $|x| = m$. Then

$$\widehat{T}(x^m) = I = \begin{bmatrix} \zeta_1^m & & 0 \\ & \ddots & \\ 0 & & \zeta_k^m \end{bmatrix},$$

so $\zeta_i^m = 1$, all i. But then $\zeta_i^{-1} = \overline{\zeta_i}$, and so

$$\widehat{T}(x^{-1}) = \begin{bmatrix} \zeta_1^{-1} & & 0 \\ & \ddots & \\ 0 & & \zeta_k^{-1} \end{bmatrix} = \begin{bmatrix} \overline{\zeta_1} & & 0 \\ & \ddots & \\ 0 & & \overline{\zeta_k} \end{bmatrix}.$$

Hence $\chi(x^{-1}) = \overline{\zeta_1} + \overline{\zeta_2} + \cdots + \overline{\zeta_k} = \overline{\zeta_1 + \cdots + \zeta_k} = \overline{\chi(x)}.$ $\triangle$

Remark It is *not* necessary to assume in Proposition 5.2.5 that F is closed under complex conjugation.

Exercise

Suppose that $F \subseteq \mathbb{C}$, χ is an F-character of G, and $z \in G$ is an involution, i.e. an element of order 2. Show that $\chi(z) \in \mathbb{Z}$, and that $\chi(z) \equiv \chi(1)$ (mod 2).

We will assume until further notice that $F \subseteq \mathbb{C}$.

If φ and θ are functions from G to F define

$$(\varphi, \theta) = |G|^{-1} \sum \{\varphi(x)\theta(x^{-1}) : x \in G\}.$$

It is easy to verify that this defines a symmetric bilinear form on the vector space of all such functions.

Remarks If χ is an F-character of G then

$$(\chi, \chi) = |G|^{-1} \sum_x \chi(x)\overline{\chi(x)} > 0$$

(by Proposition 5.2.5). Furthermore, if ψ is another F-character then $(\chi, \psi) \in \mathbb{R}$, for

$$\begin{aligned} \overline{(\chi, \psi)} &= |G|^{-1} \sum_x \overline{\chi(x)\psi(x^{-1})} \\ &= |G|^{-1} \sum_x \chi(x^{-1})\psi(x) = (\psi, \chi) = (\chi, \psi). \end{aligned}$$

Theorem 5.2.6 (1) *If χ and ψ are distinct irreducible F-characters then* $(\chi, \psi) = 0$. (2) *If χ is absolutely irreducible then* $(\chi, \chi) = 1$.

Proof (1) Say that $\chi = \chi_T$ and $\psi = \psi_S$. Choose bases so that $\hat{T} = [t_{ij}]$ and $\hat{S} = [s_{ij}]$. Then

$$
\begin{aligned}
(\chi, \psi) &= |G|^{-1} \sum_x \chi(x)\psi(x^{-1}) = |G|^{-1} \sum_x \sum_{i,j} t_{ii}(x)s_{jj}(x^{-1}) \\
&= |G|^{-1} \sum_{i,j} \sum_x t_{ii}(x)s_{jj}(x^{-1}) = 0
\end{aligned}
$$

by Theorem 5.1.8.

(2) Just as in the proof of part 1 we have

$$
\begin{aligned}
(\chi, \chi) &= |G|^{-1} \sum_x \sum_{i,j} t_{ii}(x)t_{jj}(x^{-1}) \\
&= |G|^{-1} \sum_{i,j} \frac{\delta_{ij}|G|}{\deg(T)} \\
&= \sum_i \frac{1}{\deg(T)} = \frac{\deg(T)}{\deg(T)} = 1,
\end{aligned}
$$

again using Theorem 5.1.8.   △

Corollary 5.2.7 *Any set* $\{\chi_1, \ldots, \chi_k\}$ *of distinct irreducible F-characters of G is F-linearly independent.*

Proof Suppose that $\sum_i \alpha_i \chi_i = 0$, $\alpha_i \in F$. Then

$$
0 = (0, \chi_j) = \left(\sum_i \alpha_i \chi_i, \chi_j\right) = \sum_i \alpha_i (\chi_i, \chi_j) = \alpha_j (\chi_j, \chi_j),
$$

so $\alpha_j = 0$, all j, since $(\chi_j, \chi_j) > 0$.   △

The next theorem is simply another corollary of Theorem 5.2.6; it is sin-gled out as a separate theorem because of the important role it plays in character theory.

Theorem 5.2.8 (The First Orthogonality Relation) *If $F \subseteq \mathbb{C}$ is a splitting field for G, and if $\chi_1, \ldots, \chi_k$ are all the (absolutely) irreducible F-characters of G, then*

$$
(\chi_i, \chi_j) = \delta_{ij}
$$

for all i, j.

The following two corollaries to Theorem 5.2.8 are often referred to as the "Fourier analysis of finite groups." Continue to assume, as in the theorem, that $F \subseteq \mathbb{C}$ is a splitting field for G.

Corollary 5.2.9 *If* $\chi_1, \ldots, \chi_k$ *are all the absolutely irreducible characters of* G, *and* χ *is any* F-*character, then*

$$\chi = \sum_{i=1}^{k} (\chi, \chi_i) \chi_i.$$

Proof It follows from Corollary 5.2.4 that $\chi = \sum_{i=1}^{k} n_i \chi_i$, where $0 \le n_i \in \mathbb{Z}$, all i. Thus $(\chi, \chi_j) = (\sum_i n_i \chi_i, \chi_j) = \sum_i n_i (\chi_i, \chi_j) = n_j$, all j. $\triangle$

Those absolutely irreducible χ_i for which $n_i = (\chi, \chi_i) > 0$ are called the *constituents* of χ, and the integers $n_i = (\chi, \chi_i)$ are the *multiplicities* of the constituents.

Corollary 5.2.10 *If* $\chi_1, \ldots, \chi_k$ *are all the absolutely irreducible characters of* G, *and if* $\chi = \sum_{i=1}^{k} m_i \chi_i$ *and* $\psi = \sum_{i=1}^{k} n_i \chi_i$ *are any two* F-*characters of* G, *then* $(\chi, \psi) = \sum_i m_i n_i \ge 0$ *in* $\mathbb{Z}$. *In particular,* $(\chi, \chi) = \sum_i m_i^2$.

Proof $(\chi, \psi) = (\sum_i m_i \chi_i, \sum_j n_j \chi_j) = \sum_{i,j} m_i n_j (\chi_i, \chi_j) = \sum_i m_i n_i.$ $\triangle$

Theorem 5.2.11 *If* χ *is an* F-*character of* G *then* χ *is absolutely irreducible if and only if* $(\chi, \chi) = 1$.

Proof ($\Rightarrow$) Theorem 5.2.6.

($\Leftarrow$) Let $K \supseteq F$ be a splitting field for G, and say that $\chi_1, \ldots, \chi_k$ are all the (absolutely) irreducible K-characters of G. Write $\chi = \sum_i n_i \chi_i$. Then $1 = (\chi, \chi) = \sum_i n_i^2$ by Corollary 5.2.10, and so one $n_i = 1$ and all others are 0. Thus $\chi = \chi_i$ is absolutely irreducible. $\triangle$

Theorem 5.2.12 *Suppose that* S *and* T *are* F-*representations of* G, *with* $\chi_S = \chi_T$. *Then* $S \sim T$.

Proof Let $K \supseteq F$ be a splitting field for G and let $T_1, \ldots, T_k$ be a full set of inequivalent irreducible K-representations of G, with characters $\chi_1, \ldots, \chi_k$. For any $m \ge 0$ in $\mathbb{Z}$ write mT_i to denote the direct sum of m copies of T_i. By Maschke's Theorem (5.1.1) $T^K \sim m_1 T_1 \oplus \cdots \oplus m_k T_k$, and likewise $S^K \sim n_1 T_1 \oplus \cdots \oplus n_k T_k$, $0 \le m_i, n_i \in \mathbb{Z}$. Thus $\chi_S = \chi_T = \sum_i m_i \chi_i = \sum_i n_i \chi_i$, and $m_i = n_i$ for all i by Corollary 5.2.7, and $S^K \sim T^K$. But then $S \sim T$ by Proposition 5.1.6. $\triangle$

Remark The assumption that char$(F) = 0$ is essential in Theorem 5.2.12. For example, if char$(F) = 2$, G is cyclic of order 3, and if S and T are both trivial representations, but of different odd degrees, then they have the same character but are obviously inequivalent.

Suppose that G acts as a permutation group on a set $X = \{x_1, \ldots, x_n\}$. Let T be the corresponding permutation representation (see Example 4 on page 99). If θ is the character of T and $g \in G$ note that $\theta(g)$ is just the number of 1's on the diagonal of $\widehat{T}(g)$, which is the number of *fixed points* of g in X. As a special case (Example 5, page 99) we have the right regular representation R – denote its character by $\rho = \rho_G$, the *regular character* of G. Observe that

$$\rho(g) = \begin{cases} |G| & \text{if } g = 1, \\ 0 & \text{otherwise.} \end{cases}$$

Proposition 5.2.13 *If ρ is the regular character, and $\chi_1, \ldots, \chi_k$ are all the absolutely irreducible characters of G, then $\rho = \sum_{i=1}^k \chi_i(1)\chi_i$; i.e. every absolutely irreducible character of G is a constituent of ρ, with multiplicity equal to its degree.*

Proof Apply Corollary 5.2.9 to see that the multiplicity of χ_i is

$$(\rho, \chi_i) = |G|^{-1} \sum \{\rho(x)\chi_i(x^{-1}) : x \in G\} = |G|^{-1}\rho(1)\chi_i(1) = \chi_i(1).$$

$\triangle$

Theorem 5.2.14 *If $\chi_1, \chi_2, \ldots, \chi_k$ are all the absolutely irreducible characters of G then $\sum_{i=1}^k \chi_i(1)^2 = |G|$.*

Proof $|G| = \rho(1) = \sum_i \chi_i(1)\chi_i(1)$. $\triangle$

Notation Until further notice $F \subseteq \mathbb{C}$ will denote a splitting field for G, $T_1, T_2, \ldots, T_k$ a full set of inequivalent absolutely irreducible F-representations, with characters $\chi_1 = 1_G, \chi_2, \ldots, \chi_k$. Write $\mathrm{Irr}(G)$ to denote the set $\{\chi_1, \chi_2, \ldots, \chi_k\}$ of (absolutely) irreducible characters. Let $m = c(G)$ be the *class number* of G (the number of conjugacy classes), and let $K_1 = \{1\}, K_2, \ldots, K_m$ be the conjugacy classes.

The *character table* of G is the $k \times m$ matrix whose rows and columns are indexed by the absolutely irreducible characters and the conjugacy classes of G, respectively, and whose entries are the character values at the classes.

For an easy example take G to be the symmetric group S_3, with $K_1 = \{1\}$, $K_2 = \mathrm{cl}(123)$, and $K_3 = \mathrm{cl}(12)$. There are two linear characters, $\chi_1 = 1_G$ and χ_2 the alternating character (Example 3, page 99). The permutation character θ of S_3 has values $\theta(K_1) = 3$, $\theta(K_2) = 0$, and $\theta(K_3) = 1$ (count fixed points). It is not absolutely irreducible – since $(\theta, \theta) = 2$ it must have two constituents. Since $(\theta, \chi_1) = 1$, χ_1 is a constituent of θ with multiplicity 1, and hence $\chi_3 = \theta - \chi_1$ is the other constituent. The list χ_1, χ_2, χ_3 is complete by Theorem 5.2.14, because the sum of the squares of the degrees is 6. The character table is as follows.

	K_1	K_2	K_3
χ_1	1	1	1
χ_2	1	1	-1
χ_3	2	-1	0

Exercise

Find a (matrix) representation of S_3 whose character is χ_3 above. (One possible approach is to recall that S_3 can be viewed as the symmetry group of an equilateral triangle.)

Remark In calculating (χ_i, χ_j) from the character table it is necessary to take account of the sizes of the conjugacy classes. Observe in fact that

$$(\chi_i, \chi_j) = |G|^{-1} \sum_s |K_s| \chi_i(K_s) \overline{\chi_j(K_s)}.$$

This can be useful when applying the First Orthogonality Relation (5.2.8) in cases where some of the values of one of the characters are not known yet.

Proposition 5.2.15 *For each χ_i and K_j define*

$$M_{ij} = \sum \{T_i(x): x \in K_j\} \quad and \quad \omega_{ij} = \frac{|K_j| \chi_i(K_j)}{\chi_i(1)}.$$

Then $M_{ij} \in \mathcal{C}(T_i)$, and in fact $M_{ij} = \omega_{ij} 1$.

Proof If $y \in G$ then $T_i(y)^{-1} M_{ij} T_i(y) = \sum \{T_i(y^{-1}xy): x \in K_j\} = M_{ij}$, so $M_{ij} \in \mathcal{C}(T_i)$. Thus $M_{ij} = \lambda_{ij} 1$ for some $\lambda_{ij} \in F$ by Proposition 5.1.4. Now $\mathrm{tr}(M_{ij}) = \lambda_{ij} \chi_i(1) = \sum \{\chi_i(x): x \in K_j\} = |K_j| \chi_i(K_j)$, and so $\lambda_{ij} = |K_j| \chi_i(K_j)/\chi_i(1) = \omega_{ij}$. $\triangle$

Proposition 5.2.16 *If $x \in K_s$ define*

$$n_{ijs} = |\{(y, z) \in K_i \times K_j: yz = x\}|$$

for $1 \leq i, j, s \leq m$. Then

1. n_{ijs} depends only on i, j, and s, and not on the choice of $x \in K_s$;

2. $\omega_{ti} \omega_{tj} = \sum_s n_{ijs} \omega_{ts}$; and

3. $|K_i||K_j| \chi_t(K_i)\chi_t(K_j) = \chi_t(1) \sum_s n_{ijs} |K_s| \chi_t(K_s)$,

for $1 \leq t \leq k$.

Proof (1) This is easy; just conjugate x, y and z.

(2) Multiply:

$$\begin{aligned}
M_{ti} M_{tj} &= \sum \{T_t(y): y \in K_i\} \sum \{T_t(z): z \in K_j\} \\
&= \sum \{T_t(yz): (y, z) \in K_i \times K_j\} \\
&= \sum \{T_t(x): x = yz, \ (y, z) \in K_i \times K_j\}
\end{aligned}$$

$$= \sum_s n_{ijs} \sum \{T_t(x) : x \in K_s\} = \sum_s n_{ijs} M_{ts}.$$

Apply Proposition 5.2.15 to conclude that

$$\omega_{ti}\omega_{tj} = \sum_s n_{ijs}\omega_{ts}.$$

(3) This is now an easy consequence of part 2. Multiply both sides by $\chi_t(1)^2$ and recall the definition of the ω's.   △

Exercise

Show that $n_{ijk} = n_{jik}$ for all i, j, k.

Theorem 5.2.17 (The Second Orthogonality Relation) *The columns of the character table of G are orthogonal with respect to the usual inner product of complex column vectors. Explicitly*

$$\sum_{t=1}^{k} \chi_t(K_i)\overline{\chi_t(K_j)} = \frac{\delta_{ij}|G|D}{|K_i|} = \delta_{ij}|C_G(x_i)|,$$

$1 \le i, j \le m$, *where* $x_1, \ldots, x_k$ *are representative elements in the conjugacy classes.*

Proof Say that $K_j^{-1} = K_\ell$. Then, by Proposition 5.2.16,

$$|K_i||K_\ell|\chi_t(K_i)\chi_t(K_\ell) = \chi_t(1) \sum_s n_{i\ell s}|K_s|\chi_t(K_s),$$

all t. Sum over t and apply Proposition 5.2.13:

$$|K_i||K_j| \sum_t \chi_t(K_i)\chi_t(K_j^{-1}) = \sum_s n_{i\ell s}|K_s| \sum_t \chi_t(1)\chi_t(K_s)$$

$$= \sum_s n_{i\ell s}|K_s|\rho(K_s) = n_{i\ell 1}|G|,$$

where ρ is the regular character. But

$$n_{i\ell 1} = |\{(y, z) \in K_i \times K_\ell : yz = 1\}| = |K_j|\delta_{ij}$$

since $K_\ell = K_j^{-1}$. The theorem follows upon division by $|K_i||K_j|$.   △

Remark Note in particular that $\sum_t \chi_t(K_i)\overline{\chi_t(K_i)} = |G|/|K_i| = |C_G(x)|$ for any $x \in K_i$. Thus orders of centralizers and sizes of conjugacy classes are easily determined from the character table (as is $|G|$).

For example, the following is the character table of a group G.

	K_1	K_2	K_3	K_4	K_5
χ_1	1	1	1	1	1
χ_2	1	1	1	-1	-1
χ_3	2	2	-1	0	0
χ_4	3	-1	0	1	-1
χ_5	3	-1	0	-1	1

Thus we may read from the table that $|G| = 24$, and that the sizes of the conjugacy classes are, respectively, $1, 3, 8, 6,$ and 6. Does that sound familiar?

Theorem 5.2.18 *The number k of absolutely irreducible characters of G is equal to the class number m, so the character table is a square matrix.*

Proof Let G act on $X = G$ by conjugation, with permutation character θ. Thus $\theta(g) = |C_G(g)|$, all $g \in G$, and there are $m = c(G)$ orbits. Apply the Orbit Formula (3.1.2) and the Second Orthogonality Relation (5.2.17) to see that

$$
\begin{aligned}
m &= (\theta, 1_G) = |G|^{-1} \sum \{|C_G(g)|: g \in G\} \\
&= |G|^{-1} \sum_{g \in G} \sum_{t=1}^{k} \chi_t(g) \chi_t(g^{-1}) \\
&= \sum_{t=1}^{k} (\chi_t, \chi_t) = k \cdot 1 = k.
\end{aligned}
$$

$\triangle$

Corollary 5.2.19 *The set $\mathrm{Irr}(G) = \{\chi_1, \ldots, \chi_k\}$ is a basis for the space $\mathrm{cf}_F(G)$ of F-valued class functions on G.*

Proof $\mathrm{Irr}(G)$ is a linearly independent set by Corollary 5.2.7, and it is clear that the dimension of $\mathrm{cf}(G)$ is $k = c(G)$. $\triangle$

Corollary 5.2.20 *G is abelian if and only if every absolutely irreducible character χ_i is linear; i.e. $\chi_i(1) = 1$.*

Proof ($\Rightarrow$) Proposition 5.1.7.

($\Leftarrow$) If all $\chi_i(1)$ are equal to 1 then $\sum_{i=1}^{k} \chi_i(1)^2 = k = |G| = c(G)$, and G is abelian. $\triangle$

Thus it is a trivial matter to tell from its character table whether or not a group is abelian.

The next two corollaries simply observe that the "Fourier analysis" (see page 109) of characters extends to arbitrary class functions. The proofs are virtually unchanged.

Corollary 5.2.21 *If* $\mathrm{Irr}(G) = \{\chi_1, \ldots, \chi_k\}$ *and* $\psi \in \mathrm{cf}(G)$ *then*

$$\psi = \sum_{i=1}^{k} (\psi, \chi_i)\chi_i.$$

Corollary 5.2.22 *If* $\mathrm{Irr}(G) = \{\chi_1, \ldots, \chi_k\}$ *and* $\psi, \theta \in \mathrm{cf}(G)$, *and if* $\psi = \sum_{i=1}^{k} a_i\chi_i$ *and* $\theta = \sum_{i=1}^{k} b_i\chi_i$, $a_i, b_i \in F$, *then* $(\psi, \theta) = \sum_i a_i b_i$. *In particular* $(\psi, \psi) = \sum_i a_i^2$.

Recall that $\alpha \in \mathbb{C}$ is an *algebraic integer* if it is algebraic over $\mathbb{Q}$ and its (monic) minimal polynomial $m(x)$ over $\mathbb{Q}$ is in $\mathbb{Z}[x]$. Thus, for example, $\sqrt{2}$ is an algebraic integer, but $\sqrt{2}/2$ is not.

Proposition 5.2.23 *If* $\alpha \in \mathbb{C}$ *then the following are equivalent:*

1. α is an algebraic integer,

2. $\mathbb{Z}[\alpha]$ is a finitely generated abelian group,

3. α is a root of a monic polynomial $f(x) \in \mathbb{Z}[x]$.

Proof $(1) \Rightarrow (2)$ If the minimal polynomial of α is

$$m(x) = b_0 + b_1 x + \cdots + x^k, \quad b_i \in \mathbb{Z},$$

then

$$\alpha^k = -b_0 \cdot 1 - b_1 \cdot \alpha - \cdots - b_{k-1} \cdot \alpha^{k-1} \in \mathbb{Z} \cdot 1 + \cdots + \mathbb{Z} \cdot \alpha^{k-1},$$

and it follows easily that $\{1, \alpha, \ldots, \alpha^{k-1}\}$ generates $\mathbb{Z}[\alpha]$ as an abelian group.

$(2) \Rightarrow (3)$ Let $\{f_1(\alpha), f_2(\alpha), \ldots, f_m(\alpha)\}$ be a generating set for $\mathbb{Z}[\alpha]$, with each $f_i(x) \in \mathbb{Z}[x]$. Set

$$n = \max\{1 + \deg f_i(x) : 1 \leq i \leq m\}.$$

Since $\alpha^n \in \mathbb{Z}[\alpha]$, we may write $\alpha^n = \sum_{i=1}^{m} b_i f_i(\alpha)$, $b_i \in \mathbb{Z}$. Set

$$f(x) = x^n - \sum_{i=1}^{m} b_i f_i(x) \in \mathbb{Z}[x].$$

Then $f(x)$ is monic and $f(\alpha) = 0$.

$(3) \Rightarrow (1)$ Let $m(x) \in \mathbb{Q}[x]$ be the minimal polynomial of α. Then $f(x) = m(x)g(x)$ for some $g(x) \in \mathbb{Q}[x]$. Write

$$m(x) = \left(\frac{a}{b}\right)h(x) \quad \text{and} \quad g(x) = \left(\frac{c}{d}\right)k(x),$$

with $a, b, c, d \in \mathbb{Z}$ and $h(x), k(x)$ primitive polynomials in $\mathbb{Z}[x]$. We may assume that both have positive leading coefficients. By Gauss's Lemma $h(x)k(x)$ is also primitive. But then, since $bd \cdot f(x) = ac \cdot h(x)k(x)$, we have $bd = ac$, and hence $f(x) = h(x)k(x)$ in $\mathbb{Z}[x]$. As a result $h(x)$ and $k(x)$ are monic, and hence $m(x) = h(x) \in \mathbb{Z}[x]$. $\triangle$

Corollary 5.2.24 *If $\alpha, \beta \in \mathbb{C}$ are algebraic integers then $\alpha \pm \beta$ and $\alpha\beta$ are algebraic integers. In particular, the set of all algebraic integers is a subring of $\mathbb{C}$.*

Proof Let $\{\alpha_1, \ldots, \alpha_m\}$ and $\{\beta_1, \ldots, \beta_m\}$ be generating sets for $\mathbb{Z}[\alpha]$ and $\mathbb{Z}[\beta]$, respectively, as abelian groups. Then $\{\alpha_i, \beta_j\}$ is a finite generating set for $\mathbb{Z}[\alpha, \beta]$, and $\mathbb{Z}[\alpha \pm \beta]$ and $\mathbb{Z}[\alpha\beta]$ are subgroups of $\mathbb{Z}[\alpha, \beta]$, so they are finitely generated. $\triangle$

Corollary 5.2.25 *If $\alpha \in \mathbb{Q}$, and α is an algebraic integer, then $\alpha \in \mathbb{Z}$.*

Proof The minimal polynomial over $\mathbb{Q}$ of α is $m(x) = x - \alpha$, and $m(x) \in \mathbb{Z}[x]$, so $\alpha \in \mathbb{Z}$. $\triangle$

Corollary 5.2.26 *If χ is an F-character of G, and $x \in G$, then $\chi(x)$ is an algebraic integer.*

Proof Say that $|x| = m$ and $\chi(1) = n$. See the proof of Proposition 5.2.5 to see that $\chi(x) = \zeta_1 + \cdots + \zeta_n$, where each $\zeta_i \in \mathbb{C}$ is a root of $x^m - 1$, monic in $\mathbb{Z}[x]$. Thus $\chi(x)$ is an algebraic integer by Proposition 5.2.23 and Corollary 5.2.24. $\triangle$

Recall from Proposition 5.2.15 that we defined

$$\omega_{ij} = \frac{|K_j|\chi_i(K_j)}{\chi_i(1)},$$

where $\chi_i \in \mathrm{Irr}(G)$ and K_j is a conjugacy class.

Proposition 5.2.27 *Each ω_{ij} is an algebraic integer.*

Proof Fix i and j; let v be the column vector $(\omega_{i1}, \ldots, \omega_{ik})^t$, and let N be the $k \times k$ matrix

$$\begin{bmatrix} n_{j11} & \cdots & n_{j1k} \\ & \ddots & \\ n_{jk1} & \cdots & n_{jkk} \end{bmatrix}.$$

Then $Nv = \omega_{ij}v$ by Proposition 5.2.16, so ω_{ij} is an eigenvalue of N. Thus ω_{ij} is a root of the characteristic polynomial of N, and hence is an algebraic integer by Proposition 5.2.23, since N has entries in $\mathbb{Z}$. $\triangle$

Theorem 5.2.28 *The degrees $\chi_i(1)$ of the absolutely irreducible characters of G are all divisors of $|G|$.*

Proof For each i we have

$$\frac{|G|}{\chi_i(1)} = \frac{|G|}{\chi_i(1)}(\chi_i, \chi_i)$$

$$= \sum_j \frac{|K_j|\chi_i(K_j)}{\chi_i(1)}\chi_i(K_j^{-1}) = \sum_j \omega_{ij}\chi_i(K_j^{-1}),$$

which is an algebraic integer by Corollary 5.2.26 and Proposition 5.2.27. Thus $|G|/\chi_i(1) \in \mathbb{Z}$ by Corollary 5.2.25.   △

Remark For the next few paragraphs it is sufficient to assume only that $F \subseteq \mathbb{C}$.

If $\chi = \chi_T$ define the *kernel* of χ to be $\ker(\chi) = \ker(T)$; it is well defined by Theorem 5.2.12.

Suppose that $H \triangleleft G$ and T is an F-representation of G/H on V with character χ. Write $\eta: G \to G/H$ for the usual quotient map. Then the composite map $G \xrightarrow{\eta} G/H \xrightarrow{T} \mathrm{GL}(V)$ determines a representation T' of G, viz. $T'(x) = T(xH)$ for all $x \in G$. Clearly $H \leq \ker(T')$, and if χ' is the character of T' then $\chi'(x) = \chi(xH)$, all $x \in G$. We say that T and χ have been *lifted* from G/H to G.

Conversely, if T' is an F-representation of G, with character χ' and with $H \leq \ker(T')$, then we determine a representation T of G/H via $T(xH) = T'(x)$, well defined since $H \leq \ker(T')$, and again $\chi'(x) = \chi(xH)$, all $x \in G$. It is clear in both cases that χ is irreducible if and only if χ' is irreducible.

Thus there is a 1–1 correspondence between the set of F-characters of G having H in their kernels and the set of all F-characters of G/H. It is customary to write $\chi' = \chi$ and let the context determine whether it is a character of G or of G/H.

Proposition 5.2.29 *If χ is an F-character of G then*

$$\ker(\chi) = \{x \in G: \chi(x) = \chi(1)\}.$$

Proof Say that $\chi = \chi_T$. If $x \in \ker(\chi)$ then $T(x) = 1$, so $\chi(x) = \deg(T) = \chi(1)$. Suppose conversely that $\chi(x) = \chi(1)$. As in the proof of Corollary 5.2.26 we have $\chi(x) = \zeta_1 + \cdots + \zeta_n$, $n = \chi(1)$, where each $\zeta_i \in \mathbb{C}$ is a root of unity. Thus

$$\chi(x) = \chi(1) = |\zeta_1 + \cdots + \zeta_n| = |\zeta_1| + \cdots + |\zeta_n|;$$

i.e. equality holds in the triangle inequality in $\mathbb{C}$. Thus each ζ_i is a positive multiple of ζ_1, and hence $\zeta_i = \zeta_1$ since all have absolute value 1. But then $n = n\zeta_1$, so $\zeta_1 = 1$, $\widehat{T}(x) = I$, and so $x \in \ker(T) = \ker(\chi)$.   △

Remark It follows from the proof above that $|\chi(x)| \leq \chi(1)$ for all $x \in G$.

Exercises

1. Calculate the character tables of the alternating group A_4 and the dihedral group D_4 of order 8. (*Hint* A_4 has a normal subgroup K of order 4. Obtain three linear characters for A_4/K as in Example 2, page 99, and lift them to linear characters of A_4 by composition with the quotient map $A_4 \to A_4/K$. Similar remarks apply to D_4, except that the quotient is Klein's 4-group.)

2. Find the character table of Q_2, the quaternion group of order 8, and compare it with the table of D_4.

3. Find the character table of the dihedral group

$$D_5 = \langle a, b \,|\, a^5 = b^2 = 1, ab = ba^{-1} \rangle.$$

   (*Hint* You will need the quadratic irrational $\alpha = \cos 2\pi/5$, which is equal to $(-1 + \sqrt{5})/4$. To see this use the fact that $\zeta = e^{2\pi i/5}$ is a root of

$$x^4 + x^3 + x^2 + x + 1 = x^2 \left(x^2 + x + 1 + \frac{1}{x} + \frac{1}{x^2} \right) = 0.$$

   Make the substitution $y = x + 1/x$, solve for y, then for x, and hence find α explicitly.)

4. If $p \in \mathbb{Z}$ is a prime and G is a nonabelian group of order p^3 show that G must have p^2 linear characters and $p - 1$ that are nonlinear (and irreducible), each of degree p. Conclude that $c(G) = p^2 + p - 1$ (see page 64).

Resume at this point the standing assumption that $F \subseteq \mathbb{C}$ is a splitting field for G, with corresponding notation for Irr(G), conjugacy classes, etc.

For each $\chi_i \in \mathrm{Irr}(G)$ set $N_i = \ker(\chi_i)$, so $N_i \triangleleft G$. Note that by Proposition 5.2.29 the subgroups N_i can be obtained (as unions of conjugacy classes) from the character table.

Proposition 5.2.30 *If χ is any F-character of G write $\chi = \sum_i n_i \chi_i$. Then* $\ker(\chi) = \cap \{N_i \colon n_i > 0\}$, *the intersection of the kernels of the constituents of χ.*

Proof Say that $\chi = \chi_T$, and write $J = \{i \colon 1 \leq i \leq k$ and $n_i > 0\}$. Since $T \sim \oplus \{n_i T_i \colon i \in J\}$ it is clear that $T(x) = 1$ if and only if $T_i(x) = 1$ for all $i \in J$; hence that $\ker(\chi) = \cap \{N_i \colon n_i > 0\}$. $\triangle$

Corollary 5.2.31 $\cap\{N_i: 1 \leq i \leq k\} = 1$.

Proof The right regular representation R is faithful, so $\ker(R) = \ker(\rho) = 1$. But also $\rho = \sum_i \chi_i(1)\chi_i$ by Proposition 5.2.13, so $n_i = \chi_i(1) > 0$, all i, and hence $1 = \ker(\rho) = \cap\{N_i: 1 \leq i \leq k\}$. $\triangle$

Corollary 5.2.32 *If* $H \triangleleft G$ *then* $H = \cap\{N_i: H \leq N_i\}$.

Proof $\mathrm{Irr}(G/H)$ consists of those $\chi_i \in \mathrm{Irr}(G)$ for which $H \leq N_i = \ker(\chi_i)$. By the preceding corollary the intersection of their kernels is the identity element of G/H, i.e. is H. $\triangle$

Corollary 5.2.33 *G is simple if and only if $N_i = 1$ for $2 \leq i \leq k$.*

Note that by the corollaries above all normal subgroups of G can be determined from the character table, and in particular it can be determined whether or not G is simple.

Proposition 5.2.34 *The set of linear characters of G is just $\mathrm{Irr}(G/G')$, where G' is the derived group, or commutator subgroup, of G. Furthermore $G' = \cap\{N_i: \chi_i(1) = 1\}$, so G' is determined by the character table.*

Proof Since G/G' is abelian each $\chi_i \in \mathrm{Irr}(G/G')$ is linear by Corollary 5.2.20. On the other hand, if $\chi \in \mathrm{Irr}(G)$ is linear then $\chi: G \to F^*$ is a homomorphism, and $G' \leq \ker(\chi)$ since F^* is abelian, so $\chi \in \mathrm{Irr}(G/G')$. In particular, $N_i \geq G'$ if and only if χ_i is linear, so $G' = \cap\{N_i: \chi_i(1) = 1\}$ by Corollary 5.2.32. $\triangle$

To illustrate some of the ideas above, we consider the group G whose character table appears on page 113. It has a normal subgroup $H = N_3 = \ker(\chi_3) = K_1 \cup K_2$ of order 4. The absolutely irreducible characters of G/H are just $\chi_1, \chi_2,$ and χ_3. When χ_4 and χ_5 are omitted from the table it appears as follows.

	K_1	K_2	K_3	K_4	K_5
χ_1	1	1	1	1	1
χ_2	1	1	1	-1	-1
χ_3	2	2	-1	0	0

Observe that columns 1 and 2 are identical, as are columns 4 and 5. That simply reflects the fact that classes K_1 and K_2 have merged to form a single class at the level of G/H (in fact the trivial class of G/H), and similarly for classes K_4 and K_5. Thus we may simply omit the repeated columns (and relabel slightly) to obtain the table of G/H.

	K_1	K_2	K_3
χ_1	1	1	1
χ_2	1	1	-1
χ_3	2	-1	0

Note that it is the same as the character table of S_3 derived on page 110.

The ideas in the proof of Proposition 5.2.29 suggest another useful subgroup of G associated with an F-character χ. Define

$$Z(\chi) = \{x \in G : |\chi(x)| = \chi(1)\}.$$

Clearly $N(\chi) \subseteq Z(\chi)$. A slight variation on the proof of Proposition 5.2.29 shows easily that $Z(\chi)$ consists precisely of those elements x of G that are represented by scalar multiples of the identity transformation for any representation T whose character is χ. Two immediate consequences of that description are recorded in the next proposition.

Proposition 5.2.35 *If χ is an F-character of G then $Z(\chi) \triangleleft G$. If χ is faithful then $Z(\chi) \leq Z(G)$.*

For each $\chi_i \in \mathrm{Irr}(G)$ write $Z_i = Z(\chi_i)$.

Exercise

For each $\chi_i \in \mathrm{Irr}(G)$ show that $Z_i/N_i = Z(G/N_i)$. If χ_i is faithful conclude that $Z_i = Z(G)$.

Proposition 5.2.36 $Z(G) = \cap\{Z_i : 1 \leq i \leq k\}$.

Proof Since $Z(G)N_i/N_i \leq Z(G/N_i) = Z_i/N_i$ (by the exercise above) we have $Z(G) \leq Z_i$, all i, and so $Z(G) \leq \cap_i Z_i$. Conversely, if $x \in Z_i$, all i, then $xN_i \in Z(G/N_i)$ (also by the exercise), so for any $y \in G$ we have $[xN_i, yN_i] = [x, y]N_i = N_i$ (commutator in G/N_i). Thus $[x, y] \in \cap_i N_i = 1$ (Corollary 5.2.31), and hence $x \in Z(G)$. $\triangle$

Thus we see that $Z(G)$ can be determined from the character table, and consequently the ascending central series can be determined, using first the character table of G, then that of $G/Z(G)$, and so on. In particular, it can be determined whether or not G is nilpotent.

Note that there is another easy way to determine $Z(G)$ from the character table, since $x \in Z(G)$ if and only if $|\mathrm{cl}(x)| = 1$, and the size of each class is determined by the inner product of its column with itself (see the remark following the Second Orthogonality Relation (5.2.17)).

Proposition 5.2.37 *Suppose that $\chi_i(1)$ and $|K_j|$ are relatively prime for some i, j. Then either $\chi_i(K_j) = 0$ or else $|\chi_i(K_j)| = \chi_i(1)$, in which case $K_j \subseteq Z_i$.*

Proof Choose $a, b \in \mathbb{Z}$ for which $a\chi_i(1) + b|K_j| = 1$. Multiply through by $\chi_i(K_j)/\chi_i(1)$ to obtain

$$a\chi_i(K_j) + \frac{b|K_j|\chi(K_j)}{\chi_i(1)} = \frac{\chi_i(K_j)}{\chi_i(1)},$$

or

$$a\chi_i(K_j) + b\omega_{ij} = \frac{\chi_i(K_j)}{\chi_i(1)}.$$

Thus $\chi_i(K_j)/\chi_i(1)$ is an algebraic integer by Corollary 5.2.26 and Proposition 5.2.27, and $|\chi_i(K_j)/\chi_i(1)| \leq 1$, as has been observed earlier (page 116). If the elements of K_j are of order n let $\zeta \in \mathbb{C}$ be a primitive nth root of unity, and let $\mathcal{G}$ be the Galois group of $\mathbb{Q}(\zeta)$ over $\mathbb{Q}$. Since $\chi_i(K_j)$ is a sum of powers of ζ the same is true of $\sigma(\chi_i(K_j))$ for all $\sigma \in \mathcal{G}$. Thus also $\sigma(\chi_i(K_j)/\chi_i(1))$ is an algebraic integer and $|\sigma(\chi_i(K_j)/\chi_i(1))| \leq 1$. Set

$$\alpha = \prod \left\{ \sigma\left(\frac{\chi_i(K_j)}{\chi_i(1)}\right) : \sigma \in \mathcal{G} \right\}.$$

Then α is an algebraic integer, $|\alpha| \leq 1$, and, since it is clearly fixed by all $\sigma \in \mathcal{G}$, α is in $\mathbb{Q}$. But then $\alpha \in \mathbb{Z}$ by Corollary 5.2.25, so $\alpha = 0$ or ± 1. If $\alpha = 0$ then $\chi_i(K_j) = 0$, whereas if $\alpha = \pm 1$ then every factor of α must have absolute value 1. In particular $|\chi_i(K_j)| = \chi_i(1)$, and hence $K_j \subseteq Z_i$. $\triangle$

Theorem 5.2.38 (Burnside) *If G is a nonabelian simple group then the only conjugacy class of G that has cardinality a prime power is $K_1 = \{1\}$.*

Proof Suppose that p is a prime, $|K_j| = p^m$, and $K_j \neq K_1$. Note that $m \geq 1$, since $Z(G) = 1$. Relabel the nonprincipal irreducible characters so that $p \nmid \chi_i(1)$ if $i \leq k_0$, but $p \mid \chi_i(1)$ if $i > k_0$. Since G is simple $Z_i = 1$ for all $i > 1$. By Proposition 5.2.37 it follows that $\chi_i(K_j) = 0$ for $2 \leq i \leq k_0$. Say that $\chi_i(1) = pn_i$ for each $i > k_0$. Apply the Second Orthogonality Relation (5.2.17) to columns j and 1 of the character table to obtain

$$0 = 1 + p \sum \{n_i \chi_i(K_j) : k_0 < i \leq k\}$$

(in particular $k_0 < k$, for otherwise $0 = 1$!). But then

$$\alpha = \sum_{i > k_0} n_i \chi_i(K_j) = -\frac{1}{p}$$

is an algebraic integer, contradicting Corollary 5.2.25. $\triangle$

The final theorem in this section is our first powerful application of character theoretical methods to finite group theory. It was proved by Burnside in 1904, and the first "group theoretical" (i.e. not character theoretical) proof (for odd primes) was given by D. Goldschmidt ([26]) in 1970 (see also [3]).

Theorem 5.2.39 (Burnside's $p^a q^b$-Theorem) *If p and q are primes in $\mathbb{N}$ and G is a group of order $p^a q^b$ then G is solvable.*

Proof We may assume $p \neq q$ since p-groups are solvable. If G is a minimal counterexample then it must be simple, for otherwise it would have either a smaller normal subgroup or quotient group with the same property. Let P be a Sylow p-subgroup of G, choose $x \neq 1$ in $Z(P)$, and write K for the conjugacy class (in G) of x. Then $P \leq C_G(x)$, so $|K| = [G : C_G(x)]$ is a divisor of $[G : P] = q^b$, and $|K|$ is a prime power, contradicting Theorem 5.2.38 $\triangle$

Corollary 5.2.40 *The order of a finite nonabelian simple group has at least three distinct prime divisors.*

5.3 Contragredients and Products

In this section we may assume once again that F is an arbitrary field and impose further conditions as they are needed.

If V is an F-vector space recall (e.g. see [34]) that the *dual space* is defined to be $V^\star = \text{Hom}_F(V, F)$, the space of *linear functionals* from V to F. If $\{v_1, \ldots, v_n\}$ is a basis for V then $V^\star$ has a corresponding *dual basis* $\{v_1^\star, \ldots, v_n^\star\}$, characterized by $v_i^\star(v_j) = \delta_{ij}$ for all i and j.

If $T : V \to V$ is a linear transformation define its *adjoint* transformation $T^\star : V^\star \to V^\star$ via $T^\star(\varphi) = \varphi \circ T$ (composition of functions) for all $\varphi \in V^\star$. If $S : V \to V$ is another linear transformation then $(TS)^\star \varphi = \varphi \circ (TS) = (\varphi \circ T) \circ S = (T^\star \varphi) \circ S = S^\star(T^\star \varphi) = (S^\star T^\star)\varphi$, so $(TS)^\star = S^\star T^\star$.

Suppose that T is represented by a matrix $A = [a_{ij}]$ relative to a given basis $\{v_1, \ldots, v_n\}$, i.e. $Tv_j = \sum_i a_{ij} v_i$, all j. Then

$$T^\star v_l^\star : v_j \mapsto v_l^\star(Tv_j) = v_l^\star \left(\sum_i a_{ij} v_i \right) = \sum_i a_{ij} v_l^\star(v_i) = a_{lj}$$

for each l and all j. If we define $\varphi_l = \sum_m a_{lm} v_m^\star \in V^\star$, then $\varphi_l(v_j) = \sum_m a_{lm} v_m^\star(v_j) = a_{lj}$, each l and all j. It follows that $T^\star v_l^\star = \varphi_l = \sum_m a_{lm} v_m^\star$, and hence the matrix that represents $T^\star$ relative to the dual basis is A^t, the transpose of A. In particular $\text{tr}(T^\star) = \text{tr}(T)$.

Now suppose that T is an F-representation of G on V. Define the *contragredient representation* $T^\star$ on $V^\star$ via $T^\star(x) = T(x^{-1})^\star$ for all $x \in G$. Observe that $T^\star(xy) = T(y^{-1}x^{-1})^\star = [T(y^{-1})T(x^{-1})]^\star = T^\star(x)T^\star(y)$, so $T^\star$ really is a representation of G. Note also that $\deg(T^\star) = \deg(T)$.

We summarize the above information in the next proposition.

Proposition 5.3.1 *If T is an F-represention of G on V then its contragredient $T^\star$ is an F-representation of G on the dual space $V^\star$. Relative to a basis and its dual basis we have $\widehat{T^\star}(x) = \widehat{T}(x^{-1})^t$, all $x \in G$. If χ is the character of T then the character $\chi^\star$ of $T^\star$ is given by $\chi^\star(x) = \chi(x^{-1})$.*

Corollary 5.3.2 *If $F \subseteq \mathbb{C}$ and χ is an F-character then the complex conjugate $\overline{\chi}$ is also an F-character.*

Proof By Proposition 5.2.5 $\overline{\chi}$ is the character contragredient to χ. $\triangle$

Proposition 5.3.3 *If $F \subseteq \mathbb{C}$ and if χ, ψ are F-characters of G then*

$$(\chi^{\star}, \psi^{\star}) = (\chi, \psi).$$

Proof This is a simple calculation. $\triangle$

Corollary 5.3.4 *If $F \subseteq \mathbb{C}$ then $\chi^{\star}$ is absolutely irreducible if and only if χ is absolutely irreducible. In particular, the complex conjugate of any row in the character table of G must also be a row in the table.*

Suppose that φ is a linear F-character of G and that T is an F-representation, with character χ. Then the pointwise product φT, defined by $(\varphi T)(x) = \varphi(x)T(x)$, is also an F-representation, since $\varphi(x) \in F$ for all $x \in G$. The character of φT is clearly just $\varphi\chi$, the pointwise product of functions from G to F. If $F \subseteq \mathbb{C}$ an easy calculation shows that $(\varphi\chi, \varphi\chi) = (\chi, \chi)$, so $\varphi\chi$ is absolutely irreducible if and only if χ is.

In particular, if φ, θ are two linear F-characters of G then $\varphi\theta$ is also a linear F-character. If $\varphi^{\star}$ is the contragredient of φ then $\varphi^{\star}\varphi = 1_G$, the principal character.

The next proposition should now be clear.

Proposition 5.3.5 *The set of all linear F-characters of G is an abelian group under pointwise multiplication; it acts via multiplication as a permutation group on the set of all F-characters.*

Suppose now that S and T are F-representations of G on V and W, with respective characters χ and ψ. Then $S(x) \otimes T(x)$ is a linear transformation on $V \otimes W$ for each $x \in G$, with

$$(S(x) \otimes T(x))(\sum_i v_i \otimes w_i) = \sum_i S(x)v_i \otimes T(x)w_i.$$

It is easy to check that $x \mapsto S(x) \otimes T(x)$ is an F-representation of G on $V \otimes W$; we denote it by $S \otimes T$.

If $\{v_i\}$, $\{w_j\}$ are bases for V and W then $\{v_i \otimes w_j\}$ is a basis for $V \otimes W$, so $\deg(S \otimes T) = \deg(S) \cdot \deg(T)$. Relative to the bases mentioned (with a natural ordering of $\{v_i \otimes w_j\}$) the matrix representations are related by

$$\widehat{S \otimes T}(x) = [s_{ij}(x)\widehat{T}(x)]$$

in block form, where $\widehat{S}(x) = [s_{ij}(x)]$. Consequently the character of $S \otimes T$ is $\chi\psi$, pointwise product of functions.

Again we summarize.

Proposition 5.3.6 *If χ and ψ are F-characters of G then the product $\chi\psi$ is also an F-character.*

Remark If χ and ψ are nonlinear irreducible characters then $\chi\psi$ is very often reducible. For example, if $F \subseteq \mathbb{C}$, χ is absolutely irreducible but not linear, and $\psi = \chi^\star$, then $(\chi\chi^\star, 1_G) = (\chi, \chi \cdot 1_G) = (\chi, \chi) = 1$, so 1_G is a constituent of $\chi\chi^\star$ and $\chi\chi^\star$ cannot be irreducible.

Example

We now have sufficient techniques available to derive quite easily the character table that first appeared on page 113. It is in fact the table of the symmetric group S_4. The classes are $K_1 = \{1\}$, $K_2 = \mathrm{cl}(12)(34)$, $K_3 = \mathrm{cl}(123)$, $K_4 = \mathrm{cl}(1234)$, and $K_5 = \mathrm{cl}(12)$. Since $S_4{}' = A_4$ there are just two linear characters, the principal character $\chi_1 = 1_G$ and the alternating character χ_2 (see page 99). Thus $22 = \sum_3^5 \chi_i(1)^2$, forcing the remaining degrees to be 2, 3, and 3. The permutation character θ of the action on $\{1,2,3,4\}$ has χ_1 as a constituent and gives $\chi_4 = \theta - \chi_1$, then $\chi_5 = \chi_2 \cdot \chi_4$. Finally χ_4^2 has degree 9, and we see by taking inner products that it has χ_1, χ_4, and χ_5 as constituents, each with multiplicity 1. Subtracting them from χ_4^2 yields χ_3.

Although it won't reappear until late in the next chapter, this seems to be the appropriate section in which to introduce the next concept, that of the *character ring* of G.

If $F \subseteq \mathbb{C}$ is a splitting field for G define $\mathrm{Char}(G)$ to be the set of all $\mathbb{Z}$-linear combinations of the characters in $\mathrm{Irr}(G)$:

$$\mathrm{Char}(G) = \Big\{ \sum_{i=1}^k n_i \chi_i : n_i \in \mathbb{Z} \Big\}.$$

It follows from Proposition 5.3.6 (and from Maschke's Theorem (5.1.1)) that $\mathrm{Char}(G)$ is a ring (with pointwise multiplication and addition), a subring of the ring of all functions from G to F. Elements of $\mathrm{Char}(G)$ are sometimes called *generalized characters*, or *virtual characters*. We call $\mathrm{Char}(G)$ the *character ring* of G; it is variously referred to in the literature as the *ring of generalized characters*, the *ring of virtual characters*, or even the *Grothendieck ring* of G.

Suppose that $\varphi = \sum_i n_i \chi_i \in \mathrm{Char}(G)$, and set $\psi = \sum \{ n_i \chi_i : n_i \geq 0 \}$ and $\theta = \sum \{ -n_i \chi_i : n_i \leq 0 \}$. Note that ψ and θ are both characters of G (provided that they are nonzero), and $\varphi = \psi - \theta$. Thus $\mathrm{Char}(G)$ can also be characterized as the set of differences of F-characters of G.

Suppose that G is the (internal) direct product of its subgroups H and K, that S is an F-representation of H with character φ, and that T is an

F-representation of K with character ψ. We may extend S and T to representations $\tilde{S}$ and $\tilde{T}$ of all of G by setting

$$\tilde{S}(hk) = S(h) \quad \text{and} \quad \tilde{T}(hk) = T(k)$$

for all $h \in H$, $k \in K$. The resulting characters of G are given by

$$\tilde{\varphi}(hk) = \varphi(h) \quad \text{and} \quad \tilde{\psi}(hk) = \psi(k).$$

By Proposition 5.3.6 the product $\tilde{\varphi}\tilde{\psi}$ is an F-character of G. Denote it by $\varphi \times \psi$; thus for each $hk \in G$ we have $(\varphi \times \psi)(hk) = \varphi(h)\psi(k)$.

Proposition 5.3.7 *Suppose that G is the direct product of its subgroups H and K, $F \subseteq \mathbb{C}$ is a splitting field for both H and K, $\mathrm{Irr}(H) = \{\varphi_1, \ldots, \varphi_r\}$, and $\mathrm{Irr}(K) = \{\psi_1, \ldots, \psi_s\}$. Then F is also a splitting field for G, and $\mathrm{Irr}(G) = \{\varphi_i \times \psi_j : 1 \leq i \leq r, 1 \leq j \leq s\}$.*

Proof An easy calculation shows that

$$(\varphi_i \times \psi_j, \varphi_m \times \psi_n)_G = (\varphi_i, \varphi_m)_H (\psi_j, \psi_n)_K = \delta_{im}\delta_{jn},$$

so each $\varphi_i \times \psi_j$ is absolutely irreducible and they are all distinct from one another. Since the sum of the squares of their degrees is $\sum_{i,j} \varphi_i(1)^2 \psi_j(1)^2 = |H||K| = |G|$, they constitute all of $\mathrm{Irr}(G)$. $\triangle$

Corollary 5.3.8 *If G is a finite abelian group and $F \subseteq \mathbb{C}$ is a splitting field for G then the group $\mathrm{Irr}(G)$ of linear F-characters of G is isomorphic with G.*

Proof It is sufficient to prove this in the case that $G = \langle x \rangle$ is cyclic, say of order m. In that case there must be a primitive mth root ζ of unity in F, and if we define $\chi(x) = \zeta$, then χ is a linear character. Its powers χ^i, $0 \leq i \leq m - 1$, are distinct linear characters, so they are all of $\mathrm{Irr}(G)$. $\triangle$

Corollary 5.3.9 *If $F \subseteq \mathbb{C}$ is a splitting field for G then the group of linear F-characters of G is isomorphic with G/G'.*

Proof Proposition 5.2.34 and the corollary above. $\triangle$

We close this section, and the chapter, with a theorem of Burnside about powers of a faithful character.

Theorem 5.3.10 *Suppose that $F \subseteq \mathbb{C}$ is a splitting field for G and ψ is a faithful (but not necessarily irreducible) F-character of G that takes on exactly m distinct values $a_1 = \psi(1), a_2, \ldots, a_m \in F$. Then every $\chi \in \mathrm{Irr}(G)$ is a constituent of one of the powers $\psi^0 = 1_G, \psi, \psi^2, \ldots, \psi^{m-1}$.*

Proof (Brauer) Set $A_j = \{x \in G : \psi(x) = a_j\}$, $1 \le j \le m$, and observe that $A_1 = \{1\}$ since ψ is faithful. Let $b_j = \sum\{\overline{\chi(x)} : x \in A_j\}$ for each j. Note that

$$(\psi^i, \chi) = |G|^{-1} \sum_{j=1}^{m} \sum \{\psi^i(x)\overline{\chi(x)} : x \in A_j\} = |G|^{-1} \sum_{j=1}^{m} a_j^i b_j.$$

If the theorem were false for χ we would have $(\psi^i, \chi) = 0$ for $0 \le i \le m - 1$, and therefore the following system of equations.

$$
\begin{array}{ccccccc}
b_1 & + & b_2 & + & \cdots & + & b_m & = & 0 \\
a_1 b_1 & + & a_2 b_2 & + & \cdots & + & a_m b_m & = & 0 \\
& & & \cdots & & & & & \\
a_1^{m-1} b_1 & + & a_2^{m-1} b_2 & + & \cdots & + & a_m^{m-1} b_m & = & 0
\end{array}
$$

The coefficient matrix is a Vandermonde matrix $[a_j^i]$ whose determinant is $\prod_{j<l} (a_l - a_j) \ne 0$. Thus all $b_i = 0$, but that contradicts the fact that $b_1 = \chi(1) > 0$! $\triangle$

The theorem is not generally very helpful for finding the character table, since it is usually not easy to decompose the characters ψ^i into their irreducible constituents. The regular character ρ might be viewed as an extreme example; every $\chi \in \mathrm{Irr}(G)$ is a constituent, but ρ can almost never be decomposed into its constituents without substantial further information.

Exercise

Let G be the generalized quaternion group

$$Q_m = \langle a, b \mid a^{2m} = 1, b^2 = a^m, ab = ba^{-1} \rangle.$$

Write the elements of G as $b^i a^j$, $0 \le i \le 1$, $0 \le j \le 2m - 1$.

1. Show that $G' = \langle a^2 \rangle$, and that G/G' is Klein's 4-group if m is even, cyclic of order 4 if m is odd.

2. Show that $c(G) = m + 3$ by finding the classes explicitly – the class sizes are $1, 1, 2, \ldots, 2, m, m$. (Also see the exercise on page 64.)

3. Let $\zeta = e^{\pi i/m} \in \mathbb{C}$. Define

$$T_j(a) = \begin{bmatrix} \zeta^j & 0 \\ 0 & \zeta^{-j} \end{bmatrix} \quad \text{and} \quad T_j(b) = \begin{bmatrix} 0 & (-1)^j \\ 1 & 0 \end{bmatrix},$$

$1 \le j \le m - 1$. Show that T_j is a representation of G, and if χ_j is its character show that $\chi_j \in \mathrm{Irr}(G)$, all j.

4. Write out the character table of Q_m.

Chapter 6

Induction and Restriction

In this chapter we explore the relations between characters of a group and the characters of its subgroups. It is perhaps fair to say that most of the power of character theory results from exploiting these relationships.

6.1 Modules

Suppose that F is a field and that G is a finite group. Denote by FG the set of all functions from G to F. With the usual addition of functions and multiplication by scalars it is clear that FG is an F-vector space. By a mild abuse of notation we may view each $x \in G$ as an element of FG – we identify x with the function whose value is 1 at x and 0 elsewhere. Clearly then every $a \in FG$ can be written uniquely as $\sum_{x \in G} a_x \cdot x$, with $a_x = a(x) \in F$. In particular the elements of G are a basis for FG, and its dimension is $|G|$.

The basis G is closed under (group) multiplication, and that multiplication extends naturally to a multiplication on all of FG via

$$
\left(\sum a_x x \right) \left(\sum b_y y \right) = \sum \{ a_x b_y xy : x, y \in G \}
$$
$$
= \sum \left\{ \left(\sum_{xy=u} a_x b_y \right) u : u \in G \right\}.
$$

Exercise

Verify that FG is a ring, in fact an F-algebra, with the multiplication defined above. Show that the multiplication is given by

$$
ab(x) = \sum \{ a(xz^{-1}) b(z) : z \in G \}
$$

for all $a, b \in FG$, all $x \in G$ (it is sometimes called the *convolution* product).

The F-algebra FG is called the *group algebra* of G over F.

If T is an F-representation of G on V we may extend the definition of T to elements of FG linearly, i.e. $T(\sum a_x x) = \sum a_x T(x)$. The resulting map $T: FG \to \mathrm{Hom}_F(V, V)$ is easily checked to be an F-algebra homomorphism, i.e. an F-*representation* of the algebra FG. The extended representation can now be used to define a (left) FG-module action on V via

$$\left(\sum a_x x\right)v = \sum a_x T(x)v$$

for all $v \in V$.

Thus each representation yields a unitary FG-module. Conversely, a unitary FG-module V is automatically an F-vector space, and if the module action is restricted to elements of $G \subset FG$ an F-representation T is determined by $T(x)v = xv$, all $x \in G$, $v \in V$.

Exercises

1. Suppose that S and T are F-representations of G on V and W, respectively.

   a. Show that S is irreducible if and only if V is a simple FG-module.

   b. Show that $S \sim T$ if and only if V and W are isomorphic as FG-modules.

2. View FG as a (left) FG-module via multiplication. Show that the resulting F-representation of G is the left regular representation L.

6.2   Induction

If $H \leq G$ we may view FH as a subalgebra of FG in a natural way – if

$$a = \sum_{y \in H} a_y y \in FH$$

we agree that $a_x = 0$ for all $x \in G \setminus H$, hence

$$a = \sum_{x \in G} a_x x \in FG.$$

Suppose now that $H \leq G$, and that T is an F-representation of H on V, so that V is a left FH-module. View FG as an $(FG\text{-}FH)$-bimodule and set

$$V^G = FG \otimes_{FH} V.$$

Then V^G is naturally a left FG-module, called the module *induced* from V. The FG-module V^G determines the *induced representation* T^G on the vector space V^G, and if χ is the character of T then the *induced character* χ^G is the character of T^G.

Example

Take $H = 1$ and $T = 1_H$ on $V = F$, so $FH \cong F$. Then

$$V^G \cong FG \otimes_F F \cong FG,$$

and the induced representation is (equivalent with) the left regular representation of G by Exercise 2 above.

Choose a (left) transversal (i.e. left coset representatives) $x_1, x_2, \ldots, x_m$ for H in G. If $H = \{h_1, h_2, \ldots\}$ then

$$G = \{x_i h_j : 1 \le i \le m, 1 \le j \le |H|\}.$$

For any $a \in FG$ set $a_{ij} = a(x_i h_j)$, so that

$$a = \sum_{i,j} a_{ij} x_i h_j = \sum_i x_i \Big(\sum_j a_{ij} h_j\Big) = \sum_i x_i b_i,$$

where $b_i = \sum_j a_{ij} h_j \in FH$. Thus

$$FG = x_1 FH + x_2 FH + \cdots + x_m FH,$$

a sum of right FH-submodules. It is in fact a direct sum, for if $\sum_i x_i b_i = 0$ then every $a_{ij} = 0$, and hence every $b_i = 0$. Thus, at least as F-vector spaces, we have

$$
\begin{aligned}
V^G \;&=\; FG \otimes_{FH} V \\[2mm]
&=\; (x_1 FH \oplus \cdots \oplus x_m FH) \otimes_{FH} V \\[2mm]
&\cong\; \oplus_{i=1}^m (x_i FH \otimes_{FH} V).
\end{aligned}
$$

Proposition 6.2.1 *If $H \le G$ and T is an F-representation of H on V then* $\deg(T^G) = [G\!:\!H]\deg(T)$.

Proof Each $x_i FH$ above is isomorphic as a right FH-module with FH via the map $x_i b \mapsto b$, all $b \in FH$, and so

$$x_i FH \otimes_{FH} V \cong FH \otimes_{FH} V \cong V$$

as vector spaces. Thus V^G is isomorphic with the direct sum of $m = [G\!:\!H]$ copies of V and the proposition follows. $\qquad\triangle$

Retain the notation above for the next few propositions.

Proposition 6.2.2 *If $\{v_1, \ldots, v_n\}$ is a basis for V then*

$$\{x_i \otimes v_j : 1 \le i \le m, \ 1 \le j \le n\}$$

is a basis for V^G.

Proof There are mn conceivably different $x_i \otimes v_j$'s, and mn is the dimension of V^G by Proposition 6.2.1, so it will suffice to show that they span V^G. If $u \in V^G$ write $u = \sum_r a_r \otimes u_r$, $a_r \in FG$, $u_r \in V$. Write $a_r = \sum_i x_i b_{ri}$, $b_{ri} \in FH$, so $u = \sum_{r,i} x_i b_{ri} \otimes u_r = \sum_{r,i} x_i \otimes b_{ri} u_r$. But $b_{ri} u_r \in V$, so write $b_{ri} u_r = \sum_j \alpha_{rij} v_j$, $\alpha_{rij} \in F$, and substitute to see finally that $u = \sum_{i,j} (\sum_r \alpha_{rij})(x_i \otimes v_j)$. $\triangle$

Now let $\widehat{T}$ be the matrix representation of H relative to the given basis $\{v_1, \ldots, v_n\}$, say $\widehat{T}(x) = [t_{ij}(x)]$, $x \in G$. Order the basis elements for V^G "along the rows," i.e. as follows:

$$x_1 \otimes v_1, \ldots, x_1 \otimes v_n; x_2 \otimes v_1, \ldots, x_2 \otimes v_n; \ldots ; x_m \otimes v_1, \ldots, x_m \otimes v_n.$$

Then
$$T^G(x)(x_j \otimes v_l) = x(x_j \otimes v_l) = xx_j \otimes v_l.$$

We may write $xx_j = x_i y$ for some i and some $y \in H$, and so

$$T^G(x)(x_j \otimes v_l) = x_i y \otimes v_l = x_i \otimes y v_l = x_i \otimes \widehat{T}(y) v_l$$
$$= x_i \otimes \sum_r t_{rl}(y) v_r = \sum_r t_{rl}(y)(x_i \otimes v_r).$$

Note, though, that $y = x_i^{-1} x x_j$, and therefore

$$T^G(x)(x_j \otimes v_l) = \sum_r t_{rl}(x_i^{-1} x x_j)(x_i \otimes v_r).$$

Thus there are (possibly) nonzero coefficients only for the basis elements in "row i"; viz. $x_i \otimes v_1, \ldots, x_i \otimes v_n$.

It is convenient to extend the definition of $\widehat{T}$ to all of G by agreeing that $\widehat{T}(x) = 0$ if $x \in G \setminus H$. (**Warning** The extended $\widehat{T}$ is usually *not* a representation of G!) With this convention the discussion above has proved the next proposition.

Proposition 6.2.3 *In block form the matrix representation of T^G is given by*

$$\widehat{T^G}(x) = [\widehat{T}(x_i^{-1} x x_j)],$$

with $\widehat{T}(x_i^{-1} x x_j)$ the ijth block, and with $\widehat{T}(z) = 0$ for any $z \notin H$.

We remark that there is only one nonzero block in each (block) row of $\widehat{T^G}(x)$, since for given i there is only one j for which $x_i^{-1} x x_j \in H$. Likewise there is only one nonzero block in each column.

Example
Take $H = 1$ and $T = 1_H$. Then $\widehat{T^G}(x)$ has $\widehat{T}(x_i^{-1} x x_j) = 1$ in entry ij if and only if $xx_j = x_i$, and we see again that T^G is the left regular representation.

Assume now that $F \subseteq \mathbb{C}$. Just as for $\widehat{T}$ it is useful, in order to calculate the induced character χ^G, to extend χ to a function $\dot{\chi}$ defined on all of G to F by agreeing that $\dot{\chi}|_H = \chi$ and $\dot{\chi}|_{G \setminus H} = 0$. Of course $\dot{\chi}$ tends *not* to be a character of G.

Proposition 6.2.4 *If* $F \subseteq \mathbb{C}$, $H \leq G$, χ *is an F-character of H, and* $\{x_1, \ldots, x_m\}$ *is a left transversal for H in G, then*

$$\chi^G(x) = \sum_{i=1}^{m} \dot{\chi}(x_i^{-1} x x_i) = |H|^{-1} \sum \{\dot{\chi}(t^{-1} x t) : t \in G\}$$

for all $x \in G$.

Proof We have

$$
\begin{aligned}
\chi^G(x) &= \operatorname{tr}(\widehat{T^G}(x)) = \operatorname{tr}[\widehat{T}(x_i^{-1} x x_j)] \\
&= \sum_{i=1}^{m} \operatorname{tr} \widehat{T}(x_i^{-1} x x_i) = \sum_{i=1}^{m} \dot{\chi}(x_i^{-1} x x_i).
\end{aligned}
$$

If $y \in H$ and $u \in G$, then $y^{-1} u y \in H$ if and only if $u \in H$, so $\dot{\chi}(y^{-1} x_i^{-1} x x_i y) = \dot{\chi}(x_i^{-1} x x_i)$ for all $y \in H$. Every $t \in G$ can be written uniquely as $t = x_i y$, $y \in H$, so

$$
\begin{aligned}
\sum_{t \in G} \dot{\chi}(t^{-1} x t) &= \sum \{\dot{\chi}(y^{-1} x_i^{-1} x x_i y) : y \in H,\ 1 \leq i \leq m\} \\
&= |H| \sum_{i=1}^{m} \dot{\chi}(x_i^{-1} x x_i) = |H| \chi^G(x).
\end{aligned}
$$

$\triangle$

Exercise

Let $G = S_3$ and $H = A_3$. Calculate the values of χ^G at $(1\,2)$ and $(1\,2\,3)$ for each $\chi \in \operatorname{Irr}(H)$. Which χ^G are irreducible? Write out $\widehat{T^G}$ for one of them.

It will prove useful to generalize the formula in Proposition 6.2.4. Recall that the space of all class functions from G to F is denoted $\operatorname{cf}(G)$, or $\operatorname{cf}_F(G)$ if we wish to emphasize the field. It is easy to check that $\operatorname{cf}(G)$ is in fact a ring (hence an F-algebra), multiplication being the pointwise product of functions.

Suppose that $H \leq G$. If $\varphi \in \operatorname{cf}(H)$ extend φ to a function $\dot{\varphi} : G \to F$ via $\dot{\varphi}|_{G \setminus H} = 0$. Then define the *induced class function* φ^G via

$$\varphi^G(x) = |H|^{-1} \sum \{\dot{\varphi}(t^{-1} x t) : t \in G\}$$

for all $x \in G$. Clearly $\varphi^G \in \mathrm{cf}(G)$.

Just as in the proof of Proposition 6.2.4 it is straightforward to verify that if $\{x_1, \ldots, x_m\}$ is a left transversal for H in G then

$$\varphi^G(x) = \sum_{i=1}^m \dot\varphi(x_i^{-1} x x_i);$$

a slight variation on the proof shows that if $\{y_1, \ldots, y_m\}$ is a *right* transversal then

$$\varphi^G(x) = \sum_{i=1}^m \dot\varphi(y_i x y_i^{-1}).$$

Proposition 6.2.5 *If $H \leq G$, $\varphi, \theta \in \mathrm{cf}(H)$, and $\alpha \in F$, then*

$$(\varphi + \theta)^G = \varphi^G + \theta^G \quad and \quad (\alpha\varphi)^G = \alpha\varphi^G.$$

Thus $\varphi \mapsto \varphi^G$ is a linear transformation from $\mathrm{cf}(H)$ to $\mathrm{cf}(G)$. Also $\varphi^G(1) = [G:H]\varphi(1)$.

Proof This is clear.   △

Proposition 6.2.6 (Transitivity of Induction) *Suppose that $K \leq H \leq G$ and $\varphi \in \mathrm{cf}(K)$. Then $(\varphi^H)^G = \varphi^G$.*

Proof If $x \in G$ then

$$
\begin{aligned}
(\varphi^H)^G(x) &= |H|^{-1} \sum_{t \in G} (\varphi^H)^\cdot (t^{-1} x t) \\
&= |H|^{-1} \sum_{t \in G} |K|^{-1} \sum_{s \in H} \dot\varphi(s^{-1} t^{-1} x t s) \\
&= |K|^{-1} |H|^{-1} \sum_{s \in H} \sum_{t \in G} \dot\varphi((ts)^{-1} x (ts)) \\
&= |K|^{-1} |H|^{-1} \sum_{s \in H} \sum_{u \in G} \dot\varphi(u^{-1} x u) \\
&= |K|^{-1} \sum_{u \in G} \dot\varphi(u^{-1} x u) = \varphi^G(x).
\end{aligned}
$$

                                                                          △

Remark If φ happens to be a character of G then Proposition 6.2.6 is a consequence of the fact that

$$(V^H)^G = FG \otimes_{FH} (FH \otimes_{FK} V) \cong FG \otimes_{FK} V = V^G.$$

Proposition 6.2.7 *Suppose that $H \leq G$, $\varphi \in \mathrm{cf}(H)$, and $\theta \in \mathrm{cf}(G)$. Then*

$$\varphi^G \cdot \theta = (\varphi \cdot \theta|_H)^G.$$

Thus the set of class functions induced from H form an ideal in the ring $\mathrm{cf}(G)$.

Proof If $x \in G$ then

$$
\begin{aligned}
(\varphi\theta|_H)^G(x) &= |H|^{-1} \sum_{t \in G} \dot{\varphi}(t^{-1}xt)(\dot{\theta}|_H)(t^{-1}xt) \\
&= |H|^{-1} \sum_{t \in G} \dot{\varphi}(t^{-1}xt)\theta(t^{-1}xt) \quad (\text{since } \dot{\varphi} = 0 \text{ off } H) \\
&= |H|^{-1} \sum_{t \in G} \dot{\varphi}(t^{-1}xt)\theta(x) \quad (\text{since } \theta \in \mathrm{cf}(G)) \\
&= \varphi^G(x)\theta(x) = (\varphi^G\theta)(x).
\end{aligned}
$$

$\triangle$

Theorem 6.2.8 (Frobenius Reciprocity) *If $H \leq G$, $\varphi \in \mathrm{cf}(H)$, and $\theta \in \mathrm{cf}(G)$, then*

$$(\varphi^G, \theta)_G = (\varphi, \theta|_H)_H.$$

Proof We calculate:

$$
\begin{aligned}
(\varphi^G, \theta) &= |G|^{-1} \sum_{x \in G} \varphi^G(x)\theta(x^{-1}) \\
&= |G|^{-1}|H|^{-1} \sum_{x,t} \dot{\varphi}(t^{-1}xt)\theta(x^{-1}) \\
&= |G|^{-1}|H|^{-1} \sum_t \sum_x \dot{\varphi}(t^{-1}xt)\theta(t^{-1}x^{-1}t) \\
&= |G|^{-1}|H|^{-1} \sum_t \sum_{y \in G} \dot{\varphi}(y)\theta(y^{-1}) \\
&= |H|^{-1} \sum_{y \in H} \varphi(y)\theta(y^{-1}) \quad (\text{since } \dot{\varphi} = 0 \text{ off } H) \\
&= (\varphi, \theta|_H)_H.
\end{aligned}
$$

$\triangle$

Corollary 6.2.9 *Suppose that $H \leq G$, $F \subseteq \mathbb{C}$ is a splitting field for both H and G, $\psi \in \mathrm{Irr}(H)$, and $\chi \in \mathrm{Irr}(G)$. Then the multiplicity of χ as a constituent of ψ^G is the same as the multiplicity of ψ as a constituent of $\chi|_H$.*

Exercise

If $A \leq G$, A is abelian, and $\chi \in \mathrm{Irr}(G)$, show that $\chi(1) \leq [G \colon A]$.

The next proposition ([39], page 64) looks rather technical, but it can be very useful for calculations of induced characters.

Proposition 6.2.10 *Suppose that $H \leq G$, $x \in G$, $K = \mathrm{cl}_G(x)$, and that $\varphi \in \mathrm{cf}(H)$. If $H \cap K = \varnothing$ then $\varphi^G(x) = 0$. If $H \cap K \neq \varnothing$ write $H \cap K = \dot{\bigcup}_{i=1}^m \mathrm{cl}_H(x_i)$. Then*

$$\varphi^G(x) = |C_G(x)| \sum_{i=1}^m \frac{\varphi(x_i)}{|C_H(x_i)|} = \frac{[G \colon H]}{|K|} \sum_{i=1}^m |\mathrm{cl}_H(x_i)| \varphi(x_i).$$

Proof If $H \cap K = \varnothing$ then $t^{-1}xt \notin H$ for all $t \in G$, and so $\varphi^G(x) = 0$. Suppose then that $H \cap K \neq \varnothing$. Fix x_i and choose $s \in G$ so that $s^{-1}xs = x_i$. Then $\{t \in G \colon t^{-1}xt = x_i\}$ is just the coset $C_G(x)s$ (verify). Thus

$$\begin{aligned} \varphi^G(x) &= |H|^{-1} \sum_{t \in G} \dot{\varphi}(t^{-1}xt) \\ &= |H|^{-1} \sum_{i=1}^m |C_G(x)| \varphi(x_i)| \, \mathrm{cl}_H(x_i)| \\ &= |C_G(x)| \sum_{i=1}^m \varphi(x_i)/|C_H(x_i)|; \end{aligned}$$

the second equality follows immediately.   △

Example

By way of illustration let us calculate the character table of the alternating group A_5. Label the classes as

$$K_1 = \{1\}, \quad K_2 = \mathrm{cl}(12)(34), \quad K_3 = \mathrm{cl}(123),$$

$$K_4 = \mathrm{cl}(12345), \quad K_5 = \mathrm{cl}(13524),$$

of respective sizes 1, 15, 20, 12, and 12. The only linear character is the principal character 1_G, since $A_5' = A_5$. The permutation character θ on $\{1, 2, 3, 4, 5\}$ has 2 constituents, one of them 1_G, so the other is (say) χ_4, with values 4, 0, 1, -1, -1. The degrees $\chi_1(1) = 1$ and $\chi_4(1) = 4$ are already sufficient to force the remaining degrees to be 3, 3, and 5 by Theorem 5.2.14. Now let $H = A_4 \leq A_5$, of index 5. Let φ be the character of H (see Exercise 1 on page 117) with values

$$\varphi(1) = \varphi(12)(34) = 1, \quad \varphi(123) = \omega, \quad \varphi(132) = \omega^2,$$

where $\omega^2 + \omega + 1 = 0$ in $\mathbb{C}$. Then $\varphi^G(1) = 5$. If $x = (12)(34) \in K_2$ then $K_2 \cap H = \mathrm{cl}_H(x)$, $|\mathrm{cl}_H(x)| = 3$, and, by Proposition 6.2.10, $\varphi^G(K_2) = 1$. Next set $x = (123) \in K_3$. Then

$$H \cap K_3 = \mathrm{cl}_H(123) \cup \mathrm{cl}_H(132),$$

with $|\mathrm{cl}_H(123)| = |\mathrm{cl}_H(132)| = 4$, so

$$\varphi^G(K_3) = (5/20)(4\omega^2 + 4\omega) = -1.$$

Since $K_4 \cap H = K_5 \cap H = \emptyset$ we have $\varphi^G(K_4) = \varphi^G(K_5) = 0$. Check that $(\varphi^G, \varphi^G) = 1$ and set $\chi_5 = \varphi^G$.

Next set $H = \langle \sigma \rangle$ of index 12, where σ is the 5-cycle (12345). Take $\zeta = e^{2\pi i/5} \in \mathbb{C}$ and define $\psi(\sigma) = \zeta$. Note that

$$H \cap K_2 = H \cap K_3 = \emptyset,$$

$$H \cap K_4 = \{\sigma, \sigma^{-1}\}, \quad \text{and} \quad H \cap K_5 = \{\sigma^2, \sigma^3\}.$$

By Proposition 6.2.10

$$\psi^G(K_4) = \zeta + \bar{\zeta} = 2\cos\frac{2\pi}{5} = 2\alpha,$$

$$\psi^G(K_5) = \zeta^2 + \zeta^3 = 4\alpha^2 - 2 = -2\alpha - 1,$$

where $\alpha = (-1+\sqrt{5})/4$ (see Exercise 3 on page 117). Thus ψ^G has values 12, 0, 0, 2α, $-2\alpha-1$. Check that $(\psi^G, \chi_4) = (\psi^G, \chi_5) = 1$, set $\chi_2 = \psi^G - \chi_4 - \chi_5$, and check that $(\chi_2, \chi_2) = 1$. The final character χ_3 is now easily obtained by the Second Orthogonality Relation, and the table is as follows.

	K_1	K_2	K_3	K_4	K_5
χ_1	1	1	1	1	1
χ_2	3	-1	0	$2\alpha + 1$	-2α
χ_3	3	-1	0	-2α	$2\alpha + 1$
χ_4	4	0	1	-1	-1
χ_5	5	1	-1	0	0

$$\alpha = \frac{-1+\sqrt{5}}{4}$$

Note that it is apparent from the table that A_5 is a simple group.

Proposition 6.2.11 *Suppose that G acts transitively on a set S, $s \in S$, and $H = G_s = \mathrm{Stab}_G(s)$. Then $(1_H)^G$ is the permutation character θ of G on S.*

Proof If $x \in G$ recall that $\mathrm{Fix}(x)$ denotes the set of fixed points of x in S. Now $(1_H)^G(x) = |H|^{-1} \sum_{t \in G} \dot{1}_H(txt^{-1})$. But $txt^{-1} \in H = G_s$ if and only if $s^{txt^{-1}} = s$ if and only if $(s^t)^x = s^t$, which is if and only if $s^t \in \mathrm{Fix}(x)$.

However, if t_1, $t_2 \in G$, then $s^{t_1} = s^{t_2}$ if and only if $t_1 t_2^{-1} \in G_s = H$, or $t_1 \in H t_2$, so each fixed point s^t, and hence each $txt^{-1} \in H$, occurs with multiplicity $|H|$. Thus $(1_H)^G(x) = |\operatorname{Fix}(x)| = \theta(x)$.   △

Recall (page 16) If G acts transitively on a set S, with $|S| > 1$, and $s \in S$, then $G_s = \operatorname{Stab}_G(s)$ also acts on S. The number r (≥ 2) of G_s-orbits in S is called the *rank* of G on S. Note, for example, that $r = 2$ if and only if G is doubly transitive on S.

Theorem 6.2.12 (Burnside) *Suppose that G acts transitively on S, with permutation character θ. Then the rank of G on S is $r = (\theta, \theta)$.*

Proof Set $H = G_s$ and note that $\theta|_H$ is the permutation character of H on S. Apply the Orbit Formula (Proposition 3.1.2), Frobenius Reciprocity (Theorem 6.2.8), and Proposition 6.2.11 to see that

$$r = (\theta|_H, 1_H)_H = (\theta, (1_H)^G)_G = (\theta, \theta).$$

          △

Corollary 6.2.13 *If G is transitive on S then it is doubly transitive if and only if $\theta = 1_G + \chi$ for some $\chi \in \operatorname{Irr}(G)$.*

Proof As observed above G is doubly transitive if and only if it has rank 2, hence if and only if $(\theta, \theta) = 2$, which means it has exactly 2 constituents, one of which is 1_G by the Orbit Formula (3.1.2).   △

Corollary 6.2.14 *If $n \geq 2$ then the symmetric group S_n has an absolutely irreducible character of degree $n-1$, as does the alternating group A_n if $n \geq 4$.*

6.3   Normal Subgroups and Clifford Theory

Suppose that $H \triangleleft G$ and that T is an F-representation of H. If $x \in G$ define T^x via $T^x(y) = T(^x y) = T(xyx^{-1})$, all $y \in H$. It is easy to check that T^x is also an F-representation of H – it is called the *conjugate* of T by x. If χ is the character of T then the character of T^x is the *conjugate character* χ^x, with $\chi^x(y) = \chi(^x y)$, all $y \in H$. Note that if $x, z \in G$ then $(\chi^x)^z = \chi^{xz}$.

More generally, given any $\varphi \in \operatorname{cf}(H)$ and $x \in G$, define the conjugate φ^x via $\varphi^x(y) = \varphi(^x y)$, and check that $\varphi^x \in \operatorname{cf}(H)$, $(\varphi^x)^z = \varphi^{xz}$, and $\varphi^x = \varphi$ if $x \in H$.

If $F \subseteq \mathbb{C}$, $\varphi, \theta \in \operatorname{cf}_F(H)$, and $x \in G$ then

$$
\begin{aligned}
(\varphi^x, \theta^x) &= |H|^{-1} \sum_{y \in H} \varphi^x(y) \theta^x(y^{-1}) \\
&= |H|^{-1} \sum_{y \in H} \varphi(xyx^{-1}) \theta(xy^{-1}x^{-1}) = (\varphi, \theta).
\end{aligned}
$$

In particular, if $\chi \in \mathrm{Irr}(H)$ then also $\chi^x \in \mathrm{Irr}(H)$.

We summarize.

Proposition 6.3.1 *If $H \triangleleft G$ then G acts by conjugation as a permutation group on $\mathrm{cf}(H)$ and on the set of F-characters of H. If $F \subseteq \mathbb{C}$ is a splitting field for H then G acts by conjugation on $\mathrm{Irr}(H)$.*

Note that H is in the kernel of the group actions mentioned in the proposition; whenever it is convenient we may view G/H as the group that acts.

If $H \triangleleft G$ and ψ is an F-character of H set $I_G(\psi) = \mathrm{Stab}_G(\psi)$ and call it the *inertia group* of ψ in G. Thus $\psi^x = \psi$ if and only if $x \in I_G(\psi)$. If $t = [G: I_G(\psi)]$ then $|\mathrm{Orb}_G(\psi)| = t$ and ψ has t distinct G-conjugates. In fact, if $\{x_1, \ldots, x_t\}$ is a (right) transversal for $I_G(\psi)$ in G then $\psi = \psi^{x_1}, \ldots, \psi^{x_t}$ are the distinct conjugates of ψ.

Exercise

Suppose that $H \triangleleft G$ and $\psi \in \mathrm{cf}(G)$, with $\psi|_H = \varphi \in \mathrm{Irr}(H)$. Show that $\varphi^G|_H = [G: H]\varphi$, $\varphi^G|_{G \setminus H} \equiv 0$, and $(\varphi^G, \varphi^G) = [G: H]$.

Proposition 6.3.2 *Suppose that $H \triangleleft G$, ψ is an irreducible F-character of H, and $\{x_1, \ldots, x_t\}$ is a transversal for $I_G(\psi)$ in G. Then*

$$(\psi^G)|_H = [I_G(\psi): H] \sum_{i=1}^{t} \psi^{x_i}.$$

Proof Take $y \in H$. Then

$$
\begin{aligned}
\psi^G(y) &= |H|^{-1} \sum_{x \in G} \dot\psi(xyx^{-1}) = |H|^{-1} \sum_{x \in G} \psi^x(y) \\
&= |H|^{-1} |I_G(\psi)| \sum_{i=1}^{t} \psi^{x_i}(y).
\end{aligned}
$$

$\triangle$

Theorem 6.3.3 (Clifford) *Suppose that $H \triangleleft G$, $F \subseteq \mathbb{C}$ is a splitting field for both G and H, $\chi \in \mathrm{Irr}(G)$, and $\psi \in \mathrm{Irr}(H)$ is a constituent of $\chi|_H$. Then*

$$\chi|_H = e \sum_{i=1}^{t} \psi^{x_i},$$

where $\{x_1, \ldots, x_t\}$ is a transversal for $I_G(\psi)$ in G and $e = (\chi|_H, \psi)$, the multiplicity of ψ as a constituent of $\chi|_H$.

Proof Set $e = (\chi|_H, \psi)$, and note that $e = (\chi, \psi^G)$ by Frobenius Reciprocity (6.2.8). By Proposition 6.3.2 $(\psi^G)|_H$ has the characters ψ^{x_i} as its constituents, each with multiplicity $[I_G(\psi) : H]$. If φ is any other irreducible F-character of H then $(\sum_i \psi^{x_i}, \varphi) = 0$, hence $0 = ((\psi^G)|_H, \varphi) = (\psi^G, \varphi^G)$ by Frobenius Reciprocity (6.2.8). But χ *is* a constituent of ψ^G, therefore *not* of φ^G, so $(\chi, \varphi^G) = (\chi|_H, \varphi) = 0$. Thus the only possible constituents of $\chi|_H$ are the various ψ^{x_i}, and we have $\chi|_H = \sum_i (\chi|_H, \psi^{x_i}) \psi^{x_i}$. But $(\chi|_H, \psi^{x_i}) = (\chi^{x_i}|_H, \psi^{x_i}) = (\chi^{x_i}|_H, \psi^{x_i}) = ((\chi|_H)^{x_i}, \psi^{x_i}) = (\chi|_H, \psi) = e$ for all i, and the proof is complete.   △

Corollary 6.3.4 *If 1_H is a constituent of $\chi|_H$ then $H \leq \ker(\chi)$.*

Proof Clearly $(1_H)^x = 1_H$ for all $x \in G$, so $I_G(1_H) = G$, $t = 1$, and $\chi|_H = e1_H$. But then $e = \chi(1)$, so $\chi|_H = \chi(1)1_H$ and $\chi(y) = \chi(1)$ for all $y \in H$, hence $H \leq \ker(\chi)$ by Proposition 5.2.29.   △

Remark The common multiplicity e of the conjugates of ψ in Clifford's theorem is called the *ramification index* of $\chi|_H$.

Exercise

Suppose that $H \triangleleft G$, $\chi \in \text{Irr}(G)$, and $\theta \in \text{Irr}(H)$, with $(\chi|_H, \theta) \neq 0$. Show that $\theta(1) \,\big|\, \chi(1)$.

Proposition 6.3.5 *Suppose that $\chi \in \text{Irr}(G)$ and $Z = Z(G)$. Then $\chi|_Z = \chi(1)\varphi$, where $\varphi \in \text{Irr}(Z)$ is linear, and if $z \in Z$, $x \in G$, then $\chi(zx) = \varphi(z)\chi(x)$.*

Proof Let φ be an absolutely irreducible constituent of $\chi|_Z$. Then $\varphi^x = \varphi$ for all $x \in G$, so $I_G(\varphi) = G$, $t = 1$, and $\chi|_Z = e\varphi$. But $\varphi(1) = 1$, so $e = \chi(1)$. If T is a representation of G with character χ then

$$\widehat{T}(z) = \begin{bmatrix} \varphi(z) & & 0 \\ & \ddots & \\ 0 & & \varphi(z) \end{bmatrix},$$

all $z \in Z$. Thus if $x \in G$ then

$$\chi(zx) = \text{tr}(\widehat{T}(z)\widehat{T}(x)) = \text{tr}(\varphi(z)\widehat{T}(x)) = \varphi(z)\chi(x).$$

△

Exercise

Suppose that $p \in \mathbb{Z}$ is a prime, G is a p-group with center Z, $[G:Z] = p^3$, and $\chi \in \mathrm{Irr}(G)$.

1. Show that $\chi|_Z = e\lambda$, where $e = \chi(1)$, $\lambda \in \mathrm{Irr}(Z)$, and $\lambda(1) = 1$.

2. Show that $(\chi, \lambda^G) = e$.

3. Conclude that $\chi(1)^2 \leq p^3$.

4. Conclude that $\chi(1) = 1$ or p.

For some examples of groups as in the exercise take G to have four generators a, b, c and d, subject to defining relations

$$a^p = b^p = c^p = d^p = (a,b) = (a,c) = (a,d) = (b,c) = 1, (b,d) = a, (c,d) = b,$$

where p is an odd prime.

Recall If $\chi_i \in \mathrm{Irr}(G)$ then

$$Z_i = Z(\chi_i) = \{x \in G : |\chi_i(x)| = \chi_i(1)\} \geq N_i = \ker(\chi_i).$$

Theorem 6.3.6 *If* $\chi_i \in \mathrm{Irr}(G)$ *then* $\chi_i(1) \,\big|\, [G:Z_i]$.

Proof (*Isaacs*) We may view χ_i as a character of G/N_i, and hence we may assume that χ_i is faithful. In that case $Z_i = Z(G) = Z$ (see the exercise on page 119). Define an equivalence relation R on G via xRy if and only if there is some $z \in Z$ such that $x \sim zy$ ($\sim$ denoting conjugacy in G). By Proposition 6.3.5 $\chi_i(zy) = \varphi(z)\chi_i(y)$ for some $\varphi \in \mathrm{Irr}(Z)$ (note that φ is faithful since χ_i is faithful). If xRy then $|\chi_i(x)| = |\chi_i(zy)| = |\varphi(z)\chi_i(y)| = |\chi_i(y)|$ since $|\varphi(z)| = 1$, i.e. $|\chi_i|$ is constant on the equivalence classes in G induced by R. Let $\mathcal{C}_1, \ldots, \mathcal{C}_r$ be the R-equivalence classes on which $\chi_i \neq 0$, and choose representatives $x_j \in \mathcal{C}_j$. Note that $x \in \mathcal{C}_j$ if and only if $x \sim zx_j$ for some $z \in Z$, i.e. $z^{-1}x = y \in \mathrm{cl}_G(x_j)$, or $x \in z\,\mathrm{cl}_G(x_j) \subseteq Z\,\mathrm{cl}_G(x_j)$. To repeat – $x \in \mathcal{C}_j$ if and only if $x = zy$, with $z \in Z$ and $y \in \mathrm{cl}_G(x_j)$. If $z_1, z_2 \in Z$ and $y_1, y_2 \in \mathrm{cl}_G(x_j)$, and if $z_1y_1 = z_2y_2$, then $\varphi(z_1)\chi_i(y_1) = \varphi(z_2)\chi_i(y_2)$, and $\chi_i(y_1) = \chi_i(y_2) \neq 0$, since $y_1 \sim y_2 \sim x_j$, so $\varphi(z_1) = \varphi(z_2)$. But then $z_1 = z_2$ since φ is faithful; hence also $y_1 = y_2$, and the representation of x as zy is unique. Thus $|\mathcal{C}_j| = |Z||\mathrm{cl}_G(x_j)|$. Label the classes in G so that $K_j = \mathrm{cl}_G(x_j)$ for all j. Then

$$
\begin{aligned}
|G| &= |G|(\chi_i, \chi_i) = \sum_{x \in G} |\chi_i(x)|^2 = \sum_{j=1}^r |\mathcal{C}_j||\chi_i(x_j)|^2 \\
&= |Z| \sum_{j=1}^r |K_j|\chi_i(K_j)\overline{\chi_i(x_j)} = |Z|\chi_i(1) \sum_{j=1}^r \omega_{ij}\chi_i(x_j^{-1})
\end{aligned}
$$

(see Proposition 5.2.15 for the definition of ω_{ij}). It follows from Corollary 5.2.26 and Proposition 5.2.27 that $[G\colon Z]/\chi_i(1) = \sum_j \omega_{ij}\chi_i(x_j^{-1})$ is both an algebraic integer and a rational number, and hence is a rational integer by Corollary 5.2.25.   △

Corollary 6.3.7 (Schur's Theorem) *If $\chi_i \in \mathrm{Irr}(G)$ then $\chi_i(1)$ is a divisor of $[G\colon Z(G)]$.*

Proof This is clear from the theorem, since $Z(G) \le Z_i$.   △

Theorem 6.3.8 *Suppose that $H \triangleleft G$, $\theta \in \mathrm{Irr}(H)$, and $L = I_G(\theta)$.*

1. If $\psi \in \mathrm{Irr}(L)$ is a constituent of θ^L then $\psi^G \in \mathrm{Irr}(G)$ and $(\psi^G, \theta^G) = (\psi|_H, \theta)$.

2. The map $\psi \mapsto \psi^G$ is a bijection between the set of absolutely irreducible constituents of θ^L and the set of absolutely irreducible constituents of θ^G.

Proof (1) Choose $\chi \in \mathrm{Irr}(G)$, a constituent of ψ^G. Then θ is a constituent of $\psi|_H$ and ψ is a constituent of $\chi|_L$, so θ is a constituent of $\chi|_H$. Say that $[G\colon L] = t$ and let $\theta_1 = \theta, \ldots, \theta_t$ be the distinct G-conjugates of θ, so $\chi|_H = e\sum_{i=1}^{t}\theta_i$ for some $e > 0$ in $\mathbb{Z}$ by Clifford's Theorem (6.3.3). Also, since $L = I_G(\theta) = I_L(\theta)$, we have $\psi|_H = e'\theta$ for some $e' \in \mathbb{Z}$, and since ψ is a constituent of $\chi|_L$ we have $e' \le e$. So

$$\chi(1) = et\theta(1) \le \psi^G(1) = t\psi(1) = e't\theta(1) \le et\theta(1),$$

and all are equal. Thus $\chi = \psi^G$, and furthermore

$$(\psi^G, \theta^G) = (\chi, \theta^G) = (\chi|_H, \theta) = e = e' = (\psi|_H, \theta).$$

(2) Let $\chi \in \mathrm{Irr}(G)$ be a constituent of θ^G, and let $\psi \in \mathrm{Irr}(L)$ be a constituent of $\chi|_L$. Since $\chi|_H = \chi|_L|_H$ has θ as a constituent we may choose ψ so that θ is a constituent of $\psi|_H$. Thus ψ is a constituent of θ^L, and $\chi = \psi^G$ by part 1. To prove the 1–1-ness suppose that ψ_1, ψ_2 in $\mathrm{Irr}(L)$ are distinct constituents of θ^L but that $\psi_1^G = \psi_2^G = \chi$. Then $(\chi|_H, \theta) = (\chi|_L|_H, \theta) \ge ((\psi_1 + \psi_2)|_H, \theta) > (\psi_1|_H, \theta)$, which by part 1 is equal to $(\psi_1^G, \theta^G) = (\chi, \theta^G) = (\chi|_H, \theta)$, a contradiction.   △

Theorem 6.3.9 (Ito, 1951) *If $A \triangleleft G$, A is abelian, and $\chi \in \mathrm{Irr}(G)$, then $\chi(1) \,\big|\, [G\colon A]$.*

Proof Choose $\theta \in \mathrm{Irr}(A)$ to be a constituent of $\chi|_A$, and set $L = I_G(\theta)$. By Theorem 6.3.8 $\chi = \psi^G$ for some $\psi \in \mathrm{Irr}(L)$, and $\psi|_A = e\theta$, with $e = \psi(1)$ since θ is linear. If $a \in A$ then $|\psi(a)| = e|\theta(a)| = e = \psi(1)$, so $A \le Z(\psi)$. Also $\psi(1) \,\big|\, [L\colon Z(\psi)]$ by Theorem 6.3.6, so $\psi(1) \,\big|\, [L\colon A]$. But then $t\psi(1) \,\big|\, t[L\colon A]$, and $t = [G\colon L]$, $t\psi(1) = \psi^G(1) = \chi(1)$, so $\chi(1) \,\big|\, [G\colon L][L\colon A] = [G\colon A]$.   △

Note in particular that the dihedral groups D_n, $n \geq 3$, and the quaternion groups Q_m, $m \geq 2$, have nonlinear absolutely irreducible characters only of degree 2, as they have abelian subgroups of index 2.

Theorem 6.3.10 (Blichfeldt) *Suppose that G has a normal abelian subgroup A and that G/A is a p-group for some prime p, $F \subseteq \mathbb{C}$ is a splitting field for G, and $\chi \in \mathrm{Irr}(G)$. Then $\chi = \lambda^G$ for some linear $\mathbb{C}$-character λ of a subgroup $H \leq G$.*

Proof There is no harm to the statement of the theorem if we assume for the proof that F has been enlarged to $\mathbb{C}$, in order to have ready access to absolutely irreducible characters of subgroups of G. Proceed by induction on $\chi(1)$. If $\chi(1) = 1$ there is nothing to prove, so assume that $\chi(1) > 1$ (hence $\chi(1)$ is a power of p by Ito's Theorem (6.3.9)). Denote by Φ the set of all linear characters φ of G for which $\varphi\chi = \chi$, clearly a subgroup of the group of all linear characters of G (in fact the *stabilizer* of χ). If $\varphi \in \Phi$ then $(\chi\overline{\chi}, \varphi) = (\chi, \chi\varphi) = (\chi, \chi) = 1$, so φ has multiplicity 1 in $\chi\overline{\chi}$. Write

$$\chi\overline{\chi} = \sum\{\varphi : \varphi \in \Phi\} + \sum_i \psi_i,$$

where the ψ_i are the other constituents (if any) of $\chi\overline{\chi}$. If any ψ_i were linear we would have $0 < (\chi\overline{\chi}, \psi_i) = (\chi, \chi\psi_i)$, forcing $(\chi, \chi\psi_i) = 1$, consequently $\chi = \chi\psi_i$, and $\psi_i \in \Phi$! Thus $\psi_i(1) > 1$ for all i. Since $\chi(1)\overline{\chi}(1)$ and all $\psi_i(1)$ are powers of p it follows that $p \mid |\Phi|$, and hence there is an element $\varphi_0 \in \Phi$ having order p. Then φ_0 is a homomorphism from G to the subgroup of order p in $\mathbb{C}^*$, and if $K = \ker(\varphi_0)$ then $[G : K] = p$.

Since both 1_G and φ_0 restrict to 1_K on K, the principal character 1_K has multiplicity at least 2 in $\chi\overline{\chi}|_K$, so $(\chi\overline{\chi}|_K, 1_K)_K = (\chi|_K, \chi|_K)_K \geq 2$ and $\chi|_K$ is reducible. Let θ be an absolutely irreducible constituent of $\chi|_K$, so $\theta(1) < \chi(1)$. Both are powers of p, so $p\theta(1) \leq \chi(1)$. By Frobenius Reciprocity (6.2.8), $(\chi, \theta^G)_G = (\chi|_K, \theta)_K \geq 1$, and therefore $\chi(1) \leq \theta^G(1) = p\theta(1)$. Thus $\chi = \theta^G$. By the induction hypothesis there is a subgroup $H \leq K$ and a linear character λ on H such that $\lambda^K = \theta$. But then, by Proposition 6.2.6,

$$\lambda^G = (\lambda^K)^G = \theta^G = \chi.$$

$\triangle$

Corollary 6.3.11 *If G is nilpotent and $\chi \in \mathrm{Irr}(G)$ then $\chi = \lambda^G$ for some linear $\mathbb{C}$-character λ of a subgroup $H \leq G$.*

Exercises

1. Prove Corollary 6.3.11. Recall that a finite group is nilpotent if and only if it is the direct product of its Sylow subgroups.

   Assume for the remaining problems that $H \triangleleft G$, and that $F \subseteq \mathbb{C}$ is a splitting field for G and H.

2. Show that $(1_H)^G = \sum\{\varphi(1)\varphi : \varphi \in \mathrm{Irr}(G/H)\}$. (*Hint*   Compare with $\rho_{G/H}$.)

3. If χ is any F-character of G show that $(1_H)^G \chi = (\chi|_H)^G$.

4. If χ, ψ are F-characters of G show that

$$(\chi|_H, \psi|_H) = \sum\{\varphi(1)(\varphi\chi, \psi) : \varphi \in \mathrm{Irr}(G/H)\}.$$

Assume further for the remaining problems that G/H is abelian and write $\widehat{G/H}$ for the group $\mathrm{Irr}(G/H)$.

5. If χ, $\psi \in \mathrm{Irr}(G)$ show that the restrictions $\chi|_H$ and $\psi|_H$ are equal if and only if $\chi \in \mathrm{Orb}_{\widehat{G/H}}(\psi)$.

6. If $\chi \in \mathrm{Irr}(G)$ show that $(\chi|_H, \chi|_H) = |\mathrm{Stab}_{\widehat{G/H}}(\chi)|$.

7. If $\chi \in \mathrm{Irr}(G)$ show that the restriction $\chi|_H$ is in $\mathrm{Irr}(H)$ if and only if $\mathrm{Stab}_{\widehat{G/H}}(\chi) = 1$.

6.4   Mackey Theorems

Suppose that $H \leq G$ and that T is an F-representation of H with character ψ. If $x \in G$ define T^x on H^x via $T^x(y^x) = T(y)$ for all $y \in H$. Then T^x is clearly an F-representation of H^x, with character ψ^x given by $\psi^x(y^x) = \psi(y)$, and if also $z \in G$ then $(T^x)^z = T^{xz}$ and $(\psi^x)^z = \psi^{xz}$ on H^{xz}. As usual we may broaden the definition of conjugate character so that it applies to arbitrary class functions: if $\varphi \in \mathrm{cf}(H)$ and $x \in G$ then $\varphi^x \in \mathrm{cf}(H^x)$ is defined by $\varphi^x(y^x) = \varphi(y)$ for $y \in H$, or equivalently $\varphi^x(z) = \varphi(^x z)$ for $z \in H^x$ (and hence $^x z \in H$).

The next two theorems, and their consequences, are due to G. Mackey [45]. They analyze quite generally the interactions between induced class functions.

It will be convenient to establish notation before stating the theorems. Suppose that K, $H \leq G$ and let $x_1, \ldots, x_s$ be a set of $(K\text{-}H)$-double coset representatives in G, with $x_1 = 1$. For each i set $K^{(i)} = K^{x_i}$ and $H_i = K^{(i)} \cap H$. If $\psi \in \mathrm{cf}(K)$ define $\psi_i \in \mathrm{cf}(K^{(i)})$ via $\psi_i = \psi^{x_i}$.

Theorem 6.4.1 *If $\psi \in \mathrm{cf}(K)$ then $(\psi^G)|_H = \sum_{i=1}^{s}(\psi_i|_{H_i})^H$.*

Proof   The subgroup H of G acts (by right multiplication) on the set $\{Kx\}$ of right cosets of K in G; for a fixed Kx_i the union of the cosets in the H-orbit of Kx_i is the double coset Kx_iH, and it is easy to check that $\mathrm{Stab}_H(Kx_i)$ is $K^{x_i} \cap H = H_i$. Let $h_{i1}, \ldots, h_{is_i}$ be a (right) transversal for H_i in H, so Kx_iH

is the disjoint union $Kx_ih_{i1}\dot\cup\cdots\dot\cup Kx_ih_{is_i}$. Thus $\{x_ih_{ij}\colon 1 \le i \le s,\, 1 \le j \le s_i\}$ is a transversal for K in G.

Make the trivial observation that if $z \in H$ then $z \in K^{(i)}$ if and only if $z \in H_i$, and hence $(\psi_i|_{H_i})\dot{}(z) = \dot\psi_i(z)$ for all i. So, if $y \in H$,

$$
\begin{aligned}
\psi^G(y) &= \sum_{i,j}\dot\psi(x_ih_{ij}yh_{ij}^{-1}x_i^{-1}) = \sum_{i,j}\dot\psi_i(h_{ij}yh_{ij}^{-1})\\
&= \sum_i\sum_j(\psi_i|_{H_i})\dot{}(h_{ij}yh_{ij}^{-1}) = \sum_i(\psi_i|_{H_i})^H(y).
\end{aligned}
$$

$\triangle$

Theorem 6.4.2 *If $F \subseteq \mathbb{C}$, $\varphi \in \mathrm{cf}(H)$, and $\psi \in \mathrm{cf}(K)$ then*

$$
(\varphi^G,\psi^G) = \sum_{i=1}^{s}(\varphi|_{H_i},\psi_i|_{H_i}).
$$

Proof We apply Frobenius Reciprocity (6.2.8) twice and the theorem above:

$$
(\varphi^G,\psi^G) = (\varphi,(\psi^G)|_H) = \sum_{i=1}^{s}(\varphi,(\psi_i|_{H_i})^H) = \sum_{i=1}^{s}(\varphi|_{H_i},\psi_i|_{H_i}). \qquad \triangle
$$

For the next theorem we specialize to the case $K = H$. If $x \in G$ define $H_{(x)}$ to be $H \cap H^x$; thus in the notation introduced above we have $H_i = H_{(x_i)}$.

Theorem 6.4.3 *Suppose that $H \le G$ and that $F \subseteq \mathbb{C}$ is a splitting field for G and its subgroups. If $\psi \in \mathrm{Irr}(H)$ then $\psi^G \in \mathrm{Irr}(G)$ if and only if $\psi|_{H_{(x)}}$ and $\psi^x|_{H_{(x)}}$ have no irreducible constituents in common for all $x \in G \setminus H$.*

Proof This follows directly from Theorem 6.4.2, with $\varphi = \psi$, and the fact that any $x \in G \setminus H$ can be taken to be one of the double coset representatives x_i, $i \ge 2$. $\triangle$

Corollary 6.4.4 *If $H \triangleleft G$ and $\psi \in \mathrm{Irr}(H)$ then $\psi^G \in \mathrm{Irr}(G)$ if and only if $\psi^x \ne \psi$ for all $x \in G \setminus H$.*

Proof Since $H \triangleleft G$, $H_{(x)} = H$, all $x \in G$. $\triangle$

Exercise

Suppose that $H \triangleleft G$, $\psi,\theta \in \mathrm{Irr}(H)$, and $\psi^G, \theta^G \in \mathrm{Irr}(G)$. Show that $\psi^G = \theta^G$ if and only if $\psi = \theta^x$ for some $x \in G$, i.e. if and only if ψ and θ are in the same G-orbit in $\mathrm{Irr}(H)$.

For a very special case of Clifford's Theorem (6.3.3) suppose that $[G\colon H] = 2$. Then $G/H \cong S_2$ has just two linear characters, viz. the principal character $1_{G/H}$ and the alternating character $\epsilon_{G/H}$, which lift, respectively, to 1_G and $\epsilon = \epsilon_G$, which is 1 on H and -1 on $G \setminus H$. Call ϵ the *alternating character* of G relative to H.

Proposition 6.4.5 *Suppose that* $[G\colon H] = 2$, *and let* $F \subseteq \mathbb{C}$ *be a splitting field for* G *and* H. *Let* ϵ *be the alternating character of* G *relative to* H, *and take* $x \in G \setminus H$. *If* $\chi \in \mathrm{Irr}(G)$ *set* $\theta = \chi|_H$. *Then either*

1. $\chi \neq \epsilon\chi$, *in which case* $\theta \in \mathrm{Irr}(H)$, $\theta = \theta^x$, *and* $\theta^G = \chi + \epsilon\chi$; *or*

2. $\chi = \epsilon\chi$, *in which case* $\theta = \psi + \psi^x$ *for some* $\psi \in \mathrm{Irr}(H)$, $\psi \neq \psi^x$, *and* $\psi^G = \chi$.

All absolutely irreducible characters of H *occur either as* θ *in part 1 or as* ψ *in part 2.*

Proof Let $\psi \in \mathrm{Irr}(H)$ be a constituent of θ. We have

$$1 = (\chi, \chi) = |G|^{-1}\Big(\sum_{h \in H} |\chi(h)|^2 + \sum_{h \in H} |\chi(hx)|^2\Big)$$
$$= \frac{1}{2}(\theta, \theta) + |G|^{-1}\sum_{h \in H} |\chi(hx)|^2,$$

or

$$(\theta, \theta) + |H|^{-1}\sum_{h \in H} |\chi(hx)|^2 = 2.$$

Thus either (1) $(\theta, \theta) = 1$, or (2) $(\theta, \theta) = 2$. In case 1 $\theta = \psi \in \mathrm{Irr}(H)$, and in the context of Clifford's Theorem $e = t = 1$, $I_G(\theta) = G$, so $\theta^x = \theta$. Also $\sum_{h \in H} |\chi(hx)|^2 \neq 0$, so χ doesn't vanish on $G \setminus H$, i.e. $\chi \neq \epsilon\chi$. Furthermore $\theta^G|_H = 2\theta$, $\theta^G|_{G \setminus H} = 0$, so $\theta^G = \chi + \epsilon\chi$.

In case 2, $(\theta, \theta) = 2$, θ has two constituents, so in Clifford's Theorem $I_G(\psi) = H$, $t = 2$, $e = 1$, $\theta = \psi + \psi^x$, and $\psi^x \neq \psi$. Also $\psi^G = \chi$ by Corollary 6.4.4 and Frobenius Reciprocity.

Why do the two cases describe all possibilities? Given any $\psi \in \mathrm{Irr}(H)$ set $\eta = \psi^G = \dot\psi + \psi^x$, so $\eta|_H = \psi + \psi^x$. Then $\eta \in \mathrm{Irr}(G)$ if and only if $\psi \neq \psi^x$, which is case 2 above, again by Corollary 6.4.4. If η splits it can only split as $\eta = \chi_1 + \chi_2$, $\chi_i \in \mathrm{Irr}(G)$, by Frobenius Reciprocity, since $\eta(1) = 2\psi(1)$, and both χ_1 and χ_2 restrict to ψ on H. Also $\eta|_{G \setminus H} \equiv 0$, so $\chi_2 = \epsilon\chi_1 \neq \chi_1$, case 1 above.   △

Proposition 6.4.5 will be useful later for calculating character tables of alternating groups.

Recall that a group G is the *semidirect product* (or *split extension*) of a subgroup A by a subgroup B if

$$A \lhd G, \quad B \leq G, \quad G = AB, \quad \text{and} \quad A \cap B = 1.$$

In that case we write $G = A \rtimes B$ (or, less often, $G = B \ltimes A$).

For example, the symmetric group S_n, $n \geq 3$, is the semidirect product $A_n \rtimes B$, where A_n is the alternating group and $B = \langle(12)\rangle$. If G is the dihedral group $D_n = \langle a, b | a^n = b^2 = (ab)^2 = 1\rangle$, then $G = A \rtimes B$, with $A = \langle a\rangle$ and $B = \langle b\rangle$.

> Assume for the remainder of this section that $G = A \rtimes B$, with A abelian, and that $F \subseteq \mathbb{C}$ is a splitting field for G and its subgroups.

Then B acts by conjugation on $\mathrm{Irr}(A)$, and if $\varphi \in \mathrm{Irr}(A)$ we'll write B_φ to denote $\mathrm{Stab}_B(\varphi)$. If T is an F-representation of B_φ on V define $\varphi \times T$ on $A \rtimes B_\varphi$ via

$$(\varphi \times T)(ab) = \varphi(a)T(b).$$

The result is an F-representation of $A \rtimes B_\varphi$, for

$$
\begin{aligned}
(\varphi \times T)((a_1 b_1)(a_2 b_2)) &= \varphi(a_1(^{b_1}a_2))T(b_1 b_2) \\
&= \varphi(a_1)\varphi^{b_1}(a_2)T(b_1)T(b_2) \\
&= \varphi(a_1)T(b_1)\varphi(a_2)T(b_2) \\
&= (\varphi \times T)(a_1 b_1)(\varphi \times T)(a_2 b_2)
\end{aligned}
$$

(we used the fact that φ is linear). If ψ is the character of T, then the character of $\varphi \times T$ is $\varphi \times \psi$, where $(\varphi \times \psi)(ab) = \varphi(a)\psi(b)$.

Proposition 6.4.6 *If* $\varphi \in \mathrm{Irr}(A)$, $\psi \in \mathrm{Irr}(B_\varphi)$ *then* $\varphi \times \psi \in \mathrm{Irr}(AB_\varphi)$.

Proof Calculate: $(\varphi \times \psi, \varphi \times \psi) = (\varphi, \varphi)(\psi, \psi) = 1$. $\triangle$

Proposition 6.4.7 *If* $\varphi \in \mathrm{Irr}(A)$ *and* $\psi \in \mathrm{Irr}(B_\varphi)$ *then* $(\varphi \times \psi)^G \in \mathrm{Irr}(G)$.

Proof Set $H = AB_\varphi$, so $\varphi \times \psi \in \mathrm{Irr}(H)$. Then by Theorem 6.4.3 it will suffice to choose $x = ab \in G \setminus H$ (i.e. $b \in B \setminus B_\varphi$) and show that $\varphi \times \psi$ and $(\varphi \times \psi)^x$ have no common constituents when restricted to $H \cap H^x = A(B_\varphi \cap B_\varphi^b)$. Even more so it will suffice to show that the restrictions to A have no common constituents. But an easy calculation shows that $(\varphi \times \psi)|_A = \psi(1)\varphi$, whereas $(\varphi \times \psi)^{ab}|_A = \psi(1)\varphi^b$, and the constituents are different since $b \notin B_\varphi$.

Alternatively, it is easy to check that $I_G(\varphi \times \psi) = AB_\varphi$, so the conclusion follows from part 1 of Theorem 6.3.8. $\triangle$

Proposition 6.4.8 *Suppose that* $\varphi, \theta \in \mathrm{Irr}(A)$, $\psi \in \mathrm{Irr}(B_\varphi)$, *and* $\eta \in \mathrm{Irr}(B_\theta)$.
1. *If* $\varphi \notin \mathrm{Orb}_B(\theta)$ *then* $(\varphi \times \psi)^G \neq (\theta \times \eta)^G$.
2. *If* $\varphi = \theta$ *but* $\psi \neq \eta$ *then* $(\varphi \times \psi)^G \neq (\varphi \times \eta)^G$.

Proof Let us set the stage to apply Theorem 6.4.2. Set $K = AB_\varphi$ and $H = AB_\theta$. A set $\{b_1, \ldots, b_s\}$ of $(B_\varphi\text{-}B_\theta)$-double coset representatives in B is also a set of $(K\text{-}H)$-double coset representatives in G. Set $K^{(i)} = K^{b_i} = AB_\varphi^{b_i}$ and $H_i = K^{(i)} \cap H = AB_i$, where $B_i = B_\varphi^{b_i} \cap B_\theta$. Then by Theorem 6.4.2,

$$((\varphi \times \psi)^G, (\theta \times \eta)^G) = \sum_{i=1}^{s}((\varphi \times \psi)|_{AB_i}, (\theta \times \eta)^{b_i}|_{AB_i})$$
$$= \sum_{i=1}^{s}(\varphi, \theta^{b_i})(\psi|_{B_i}, \eta^{b_i}|_{B_i}).$$

Thus in case 1 the inner product is 0, since $\theta^{b_i} \neq \varphi$ for all i.

In case 2 the inner product becomes

$$((\varphi \times \psi)^G, (\varphi \times \eta)^G) = \sum_{i=1}^{s}(\varphi, \varphi^{b_i})(\psi|_{B_i}, \eta^{b_i}|_{B_i}),$$

but now $(\varphi, \varphi^{b_i}) = 0$ unless $b_i \in B_\varphi$, i.e. unless $i = 1$. Furthermore the summand for $i = 1$ is $(\varphi, \varphi)(\psi, \eta) = 0$, since $\psi \neq \eta$.   △

Theorem 6.4.9 *Suppose that $G = A \rtimes B$, with A abelian. Let $\varphi_1, \ldots, \varphi_m$ be representatives of the B-orbits in $\mathrm{Irr}(A)$. For each i let $B_i = B_{\varphi_i}$, and write $\mathrm{Irr}(B_i) = \{\psi_{ij} : 1 \leq j \leq c(B_i)\}$. For each i and j set $\chi_{ij} = \varphi_i \times \psi_{ij}$. Then $\mathrm{Irr}(G) = \{\chi_{ij}^G : 1 \leq i \leq m, 1 \leq j \leq c(B_i)\}$*

Proof By Propositions 6.4.7 and 6.4.8 the characters χ_{ij}^G are all irreducible, and they are distinct, so it will suffice, by Theorem 5.2.14, to show that the sum of the squares of their degrees is $|G|$. Note that $[G : AB_i] = [B : B_i]$, and that each φ_i is linear, so $\chi_{ij}^G(1) = [B : B_i]\psi_{ij}(1)$. Thus

$$\sum_{i,j}\chi_{ij}^G(1)^2 = \sum_{i,j}[B : B_i]^2\psi_{ij}(1)^2 = \sum_i[B : B_i]^2\sum_j\psi_{ij}(1)^2$$
$$= \sum_i[B : B_i]^2|B_i| = |B|\sum_i[B : B_i] = |B|\sum_i|\mathrm{Orb}_B(\varphi_i)|$$
$$= |B||\mathrm{Irr}(A)| = |B||A| = |G|.$$

△

The procedure indicated in Theorem 6.4.9 for finding all the characters of $G = A \rtimes B$ is sometimes called the *little group method* of Mackey and Wigner.

For an easy example let us apply the method to determine the characters of the dihedral group $D_m = \langle a, b \mid a^m = b^2 = (ba)^2 = 1\rangle$, with $m = 2n$ being even (the easier case of m odd is an exercise below).

The conjugacy classes are $K_0 = \{1\}$, $K_i = \{a^i, a^{-i}\}$ for $1 \leq i \leq n - 1$, $K_n = \{a^n\}$, $K_{n+1} = \{ba^j : j \text{ even}\}$, and $K_{n+2} = \{ba^j : j \text{ odd}\}$. Set $A = \langle a \rangle$ and

$B = \langle b \rangle$. If $\zeta = e^{2\pi i/m} \in \mathbb{C}$ and if $\varphi(a) = \zeta$ then $\mathrm{Irr}(A) = \langle \varphi \rangle$; we write φ_i for φ^i, $0 \le i \le m-1$. The B-orbits in $\mathrm{Irr}(A)$ are $\{\varphi_0\}$, $\{\varphi_i, \varphi_{m-i}\}$, $1 \le i \le n-1$, and $\{\varphi_n\}$, so we may take as orbit representatives $\{\varphi_i : 0 \le i \le n\}$. The stabilizers in B are $B_0 = B_n = B$, $B_i = 1$ for $1 \le i \le n - 1$. Write $\mathrm{Irr}(B) = \{\psi_0, \psi_1\}$, with $\psi_0 = 1_B$ and $\psi_1(b) = -1$. Then there are four linear characters on G, viz. $\varphi_0 \times \psi_0$, $\varphi_0 \times \psi_1$, $\varphi_n \times \psi_0$, and $\varphi_n \times \psi_1$, which we will relabel as χ_1 through χ_4, respectively. The remainder of $\mathrm{Irr}(G)$ consists of the degree 2 characters $\chi_{j+4} = \varphi_j^G$, $1 \le j \le n - 1$, whose values are easily calculated as in Proposition 6.2.4. The character table follows.

	K_0	K_i	K_n	K_{n+1}	K_{n+2}
χ_1	1	1	1	1	1
χ_2	1	1	1	-1	-1
χ_3	1	$(-1)^i$	$(-1)^n$	1	-1
χ_4	1	$(-1)^i$	$(-1)^n$	-1	1
χ_{j+4}	2	$\zeta^{ij} + \zeta^{-ij}$	$(-1)^j \cdot 2$	0	0

Compare the table of D_{2m} (order $4m$) with that of the generalized quaternion group Q_m (exercise, page 125).

Exercises

1. Calculate the character table for the dihedral group D_m if m is odd.

2. Calculate the character table of the group
$$G = \langle a, b \mid a^7 = b^3 = 1, ba = a^2 b \rangle.$$

6.5 Brauer Theorems

This section is devoted to some powerful theorems of Richard Brauer.

We begin with a very general fact ([39], page 128) about rings of $\mathbb{Z}$-valued functions on finite sets.

Proposition 6.5.1 (Isaacs-Banaschewski) *Let X be a finite set and suppose that R is a ring of functions from X to $\mathbb{Z}$, with pointwise operations. Write 1_X for the constant function with value 1 on X. Suppose that for each $x \in X$ and each prime p there is a function $f_{x,p} \in R$ such that $p \nmid f_{x,p}(x)$. Then $1_X \in R$.*

Proof For each $x \in X$ set $I_x = \{f(x): f \in R\}$, evidently an ideal in $\mathbb{Z}$. If I_x were not all of $\mathbb{Z}$ it would be contained in some ideal (p), p a prime. But then $p \mid f(x)$, all $f \in R$, contradicting $p \nmid f_{x,p}(x)$. Thus $I_x = \mathbb{Z}$, all x, and in particular there exists $g_x \in R$ with $g_x(x) = 1$. Then $\prod_{x \in X}(1_X - g_x) = 0$ and, when the product is expanded, 1_X can be expressed as a sum of elements of R. $\triangle$

Suppose that $\mathcal{H}$ is a family of subgroups H of G so that if $H, K \in \mathcal{H}$ and $x \in G$ then $H \cap K^x \in \mathcal{H}$. Define $\mathcal{B}_G(\mathcal{H})$, the *Burnside ring* of G relative to $\mathcal{H}$, to be the additive abelian group generated by all the permutation characters $(1_H)^G$, $H \in \mathcal{H}$.

Proposition 6.5.2 *The Burnside ring $\mathcal{B}_G(\mathcal{H})$ is a subring of the character ring* Char(G).

Proof It will suffice to show that the product of any two of its generators is in $\mathcal{B}_G(\mathcal{H})$. Take $H, K \in \mathcal{H}$ and apply Propositions 6.2.6 and 6.2.7 and Theorem 6.4.1 to see that

$$
\begin{aligned}
(1_H)^G(1_K)^G &= (1_H(1_K)^G|_H)^G = ((1_K)^G|_H)^G \\
&= (\sum_i (1_{H_i})^H)^G = \sum_i (1_{H_i})^G \in \mathcal{B}_G(\mathcal{H}),
\end{aligned}
$$

since each $H_i = H \cap K^{x_i} \in \mathcal{H}$. $\triangle$

A group H is called *elementary* if $H = C \times P$, where P is a p-group for some prime p and $C = \langle c \rangle$ is cyclic of order prime to p. More generally, H is called *quasi-elementary* if $H = C \rtimes P$, with C and P as above. If it is necessary to specify the prime p we say that H is *p-elementary* or *p-quasi-elementary*.

Note that subgroups of elementary groups are elementary, and likewise subgroups of quasi-elementary groups are quasi-elementary. Thus both the family $\mathcal{E}$ of elementary subgroups of G and the family $\mathcal{Q}$ of quasi-elementary subgroups of G satisfy the requirement imposed above in the definition of a Burnside ring.

Proposition 6.5.3 *The principal character 1_G is in the ring $\mathcal{B}_G(\mathcal{Q})$.*

Proof We will apply Proposition 6.5.1 to $R = \mathcal{B}_G(\mathcal{Q})$. Given a prime p and $x \in G$ write $|x| = p^e n$, with $(n, p) = 1$, set $c = x^{p^e}$ and $C = \langle c \rangle$, cyclic of order prime to p. Let P be a Sylow p-subgroup of $N = N_G(C)$, and set $H = CP \in \mathcal{Q}$. Then let $f_{x,p}$ be the permutation character $(1_H)^G$. We need to show that $p \nmid (1_H)^G(x)$. By Proposition 6.2.11 we have $(1_H)^G(x) = |\{yH: y \in G \text{ and } xyH = yH\}|$, the number of fixed points in the action of G on cosets of H. But $xyH = yH$ if and only if $x^y \in H$, in which case $C^y \leq H$. But C contains *all* the Sylow q-subgroups of H for $q \neq p$ so it follows that $C^y = C$ and $y \in N$. Thus the cosets fixed by x are all cosets of

H in N, so $(1_H)^G(x) = (1_H)^N(x)$. If $y \in N$ then $cyH = yc^yH = yH$ (since $C \lhd N$ and $C \leq H$); that is $x^{p^e}yH = yH$. Thus the permutation effected by x on $\{yH : y \in N\}$ has order dividing p^e, and any nontrivial orbits have size divisible by p. But the total number of cosets is $[N : H]$, which is prime to p, so the number of cosets fixed by x (i.e. 1-point orbits) must also be prime to p, i.e. $p \nmid (1_H)^N(x) = f_{x,p}(x)$. $\triangle$

The next two theorems will be proved simultaneously; the corollary follows immediately.

Theorem 6.5.4 (Brauer's Induction Theorem) *If $\psi \in \mathrm{Char}(G)$ there are elementary subgroups $H_1, \cdots, H_m$ of G, linear characters $\lambda_i \in \mathrm{Irr}(H_i)$, and integers a_i so that $\psi = \sum_{i=1}^m a_i\lambda_i^G$.*

Theorem 6.5.5 (Brauer's Characterization of Characters) *If ψ is a class function on G then $\psi \in \mathrm{Char}(G)$ if and only if $\psi|_H \in \mathrm{Char}(H)$ for every elementary subgroup H of G.*

Corollary 6.5.6 *If $\psi \in \mathrm{cf}(G)$ then $\psi \in \mathrm{Irr}(G)$ if and only if*

1. $\psi|_H \in \mathrm{Char}(H)$ for every elementary subgroup H of G,

2. $(\psi, \psi) = 1$, and

3. $\psi(1) > 0$.

Proof Let $\mathcal{R}$ be the ring

$$\{\varphi \in \mathrm{cf}(G) : \varphi|_H \in \mathrm{Char}(H), \text{ all } H \in \mathcal{E}\}$$

(where $\mathcal{E}$ is the set of elementary subgroups of G), and let $\mathcal{I} = \mathcal{I}_G$ be the additive group generated by all λ^G, where λ is a linear character of some $H \in \mathcal{E}$. Evidently $\mathcal{I} \subseteq \mathrm{Char}(G) \subseteq \mathcal{R}$, and to prove both theorems it will suffice to show that $\mathcal{I} = \mathcal{R}$, since $\mathcal{I} = \mathrm{Char}(G)$ proves 6.5.4 and $\mathrm{Char}(G) = \mathcal{R}$ proves 6.5.5.

We show first that $\mathcal{I}$ is an ideal in $\mathcal{R}$. If $\varphi \in \mathcal{R}$ and $\eta \in \mathcal{I}$ write $\eta = \sum_i a_i\lambda_i^G$, with $a_i \in \mathbb{Z}$ and λ_i a linear character of $H_i \in \mathcal{E}$, all i. Then $\varphi\eta = \sum_i a_i(\varphi|_{H_i}\lambda_i)^G$ by Proposition 6.2.7. Write $\varphi|_{H_i} \in \mathrm{Char}(H_i)$ as $\varphi|_{H_i} = \sum_j b_{ij}\xi_{ij}$, with all $b_{ij} \in \mathbb{Z}$ and each $\xi_{ij} \in \mathrm{Irr}(H_i)$. Since elementary subgroups are clearly nilpotent, we have $\xi_{ij} = \mu_{ij}^{H_i}$ for some linear character μ_{ij} of a subgroup K_{ij} (also automatically elementary) of H_i by Corollary 6.3.11. Substituting and applying Propositions 6.2.6 and 6.2.7, we see that

$$\varphi\eta = \sum_{i,j} a_ib_{ij}(\lambda_i|_{K_{ij}}\mu_{ij})^G \in \mathcal{I}.$$

Now it will suffice to show that $1_G \in \mathcal{I}$. We may assume inductively that the result of Theorem 6.5.4 holds for all proper subgroups of G. Thus it will suffice to show that 1_G can be written as a $\mathbb{Z}$-linear combination of characters induced from proper subgroups of G, since then by induction each of those characters is a $\mathbb{Z}$-linear combination of characters induced from linear characters of elementary subgroups, hence so is 1_G by transitivity of induction (Proposition 6.2.6).

If we knew that $1_H \in \mathcal{I}_H$ for every quasi-elementary subgroup H of G we could conclude that $1_G \in \mathcal{I}_G$ by Proposition 6.5.3 (and the transitivity of induction), so we may assume now that $G = C \rtimes P$ is quasi-elementary. Set $N = N_G(P)$ and note that $N = (N \cap C) \times P \in \mathcal{E}$. If $N = G$ there is nothing to prove, so we assume that N is a proper subgroup. Note that $((1_N)^G, 1_G) = (1_N, 1_N) = 1$, so

$$(1_N)^G = 1_G + \sum \{a_i \chi_i \colon a_i \in \mathbb{Z}, \chi_i \in \mathrm{Irr}(G), \text{ and } i \geq 2\}.$$

Let us show that if $a_i \neq 0$ then χ_i must be nonlinear. If to the contrary χ_i were linear, then $\chi_i|_N$ has 1_N as a constituent by Frobenius Reciprocity (6.2.8), and then $N \leq K = \ker(\chi_i)$ by Corollary 6.3.4. But then, since P is a Sylow p-subgroup of K, the Frattini Argument (Proposition 4.2.1) entails that $G = NK \leq K$, contradicting the fact that K is a proper normal subgroup. Thus $\chi_i(1) > 1$. By Blichfeldt's Theorem (6.3.10) each χ_i with $a_i \neq 0$ can be written as λ_i^G for some linear character λ_i of a (necessarily proper) subgroup, and finally

$$1_G = (1_N)^G - \sum_i a_i \lambda_i^G,$$

completing the proof.   △

R. Brauer first proved the theorems above in 1947, in a paper [7] on *L*-series. The proof given here, which appeared in 1975 and is due to D. Goldschmidt and I. M. Isaacs [28], is considerably shorter than Brauer's original proof.

Applications of Brauer's theorems will appear in later chapters.

Chapter 7

Computing Character Tables

In this chapter we discuss some algorithms for calculating the character table of a finite group G. We assume throughout that G has conjugacy classes $K_1, K_2, \ldots, K_k$, that $F \subseteq \mathbb{C}$ is a splitting field for G, and that $\mathrm{Irr}(G) = \{\chi_1, \ldots, \chi_k\}$. We take $K_1 = \{1\}$ and $\chi_1 = 1_G$ as usual.

7.1 Burnside

Recall from Chapter 5 (Proposition 5.2.15) that we defined

$$\omega_{ij} = \frac{|K_j|\chi_i(K_j)}{\chi_i(1)},$$

and that if $z \in K_t$ then

$$n_{rst} = |\{(x, y) \in K_r \times K_s : xy = z\}|$$

is independent of the choice of z. From Proposition 5.2.16 we have

$$\omega_{ir}\omega_{is} = \sum_{t=1}^{k} n_{rst}\omega_{it} \qquad (*)$$

for $i, r, s = 1, \ldots, k$.

As in the proof of Proposition 5.2.27 we may interpret the equation $(*)$ as follows: for $1 \leq i \leq k$ define column vectors $v_i = (\omega_{i1}, \ldots, \omega_{ik})^t$, and for $1 \leq r \leq k$ define a $k \times k$ matrix $N_r = [n_{rst}]$, often called a *class matrix*. Then the v_i's are all eigenvectors of each N_r, and the eigenvalue corresponding to v_i is ω_{ir}.

Proposition 7.1.1 *The eigenvectors $\{v_1, \ldots, v_k\}$ are linearly independent.*

Proof Suppose that $\sum_i c_i v_i = 0$, with all $c_i \in F$. Entry by entry we see that $\sum_i c_i \omega_{ij} = 0$, all j, hence from the definition of ω_{ij} that $\sum_i (c_i/\chi_i(1))\chi_i = 0$, and so all $c_i = 0$ by Corollary 5.2.7. $\triangle$

Notation Determine a permutation $j \mapsto j'$ of $\{1, \ldots, k\}$ by means of $K_{j'} = K_j^{-1}$ for each conjugacy class K_j.

Proposition 7.1.2 *If $1 \leq i, j \leq k$ then*

$$\sum_{r=1}^{k} \frac{\omega_{ir}\omega_{jr'}}{|K_r|} = \frac{\delta_{ij}|G|}{\chi_i(1)\chi_j(1)} \; .$$

Proof Calculate:

$$\sum_r \frac{\omega_{ir}\omega_{jr'}}{|K_r|} = \sum_r \frac{|K_r|\chi_i(K_r)|K_r|\chi_j(K_r^{-1})}{\chi_i(1)\chi_j(1)}$$

$$= \frac{1}{\chi_i(1)\chi_j(1)} \sum_r |K_r|\chi_i(K_r)\chi_j(K_r^{-1})$$

$$= \frac{|G|(\chi_i, \chi_j)}{\chi_i(1)\chi_j(1)} = \frac{\delta_{ij}|G|}{\chi_i(1)\chi_j(1)} \; .$$

$\triangle$

Corollary 7.1.3 *If ω_{ir} is known for all r, then $\chi_i(1)$ is determined.*

Proof Take $j = i$ in the Proposition. $\triangle$

In [10], §223, Burnside presented a procedure for calculating character tables which, he wrote, "requires only algebraical processes" provided that the integers n_{rst} are known. Based on his description of the procedure, we sketch next an algorithm for determining the character table.

The Burnside Algorithm

1. Determine the conjugacy classes of G.

2. Calculate all n_{rst}, hence all matrices N_r.

3. Find k linearly independent eigenvectors for each N_r, normalized to have first entry 1 (since $\omega_{i1} = |K_1| = 1$). The ijth entry of v_i is ω_{ij}.

4. Use Corollary 7.1.3 to calculate $\chi_i(1)$, all i.

5. Calculate $\chi_i(K_j) = \omega_{ij}\chi_i(1)/|K_j|$, all i and j.

For a very simple example take $G = S_3$, with $K_1 = \{1\}$, $K_2 = \mathrm{cl}(1\,2\,3)$, and $K_3 = \mathrm{cl}(1\,2)$. Easy calculations yield

$$N_1 = \begin{bmatrix} 1 & 0 & 0 \\ 0 & 1 & 0 \\ 0 & 0 & 1 \end{bmatrix}, \quad N_2 = \begin{bmatrix} 0 & 1 & 0 \\ 2 & 1 & 0 \\ 0 & 0 & 2 \end{bmatrix}, \quad \text{and} \quad N_3 = \begin{bmatrix} 0 & 0 & 1 \\ 0 & 0 & 2 \\ 3 & 3 & 0 \end{bmatrix}.$$

For example, $n_{332} = 3$ since $(1\,2\,3) = (1\,2)(1\,3) = (1\,3)(2\,3) = (2\,3)(1\,2)$. The respective characteristic polynomials are

$$f_1(\lambda) = (1-\lambda)^3, \quad f_2(\lambda) = -(1+\lambda)(2-\lambda)^2, \quad \text{and} \quad f_3(\lambda) = -\lambda(3-\lambda)(3+\lambda).$$

Since N_3 has distinct eigenvalues it has linearly independent eigenvectors. When normalized to have first entry 1 they are

$$v_1 = (\omega_{11}, \omega_{12}, \omega_{13})^t = (1, 2, 3)^t \quad (\lambda = 3),$$

$$v_2 = (\omega_{21}, \omega_{22}, \omega_{23})^t = (1, 2, -3)^t \quad (\lambda = -3),$$

$$v_3 = (\omega_{31}, \omega_{32}, \omega_{33})^t = (1, -1, 0)^t \quad (\lambda = 0).$$

They are of course automatically eigenvectors for N_1 and N_2 as well.

Next Corollary 7.1.3 yields the degrees $\chi_1(1) = \chi_2(1) = 1$ and $\chi_3(1) = 2$. Finally, we calculate the remaining values, $\chi_3(K_2) = \omega_{32}\chi_3(1)/|K_2| = -1$, etc. The completed character table, as we have seen earlier, is

	K_1	K_2	K_3
χ_1	1	1	1
χ_2	1	1	-1
χ_3	2	-1	0

Exercise

Apply Burnside's algorithm to calculate the character tables of the dihedral groups D_4 and D_5.

One possible approach to calculating the integers n_{rst} is to work within the group algebra FG. For a given conjugacy class K, we may define $C = \sum \{x : x \in K\} \in FG$, i.e. C is the function from G to F defined by

$$C(x) = \begin{cases} 1 & \text{if } x \in K, \\ 0 & \text{otherwise.} \end{cases}$$

If the conjugacy classes of G are $K_1, \ldots, K_k$, as usual, we obtain corresponding $C_1, \ldots, C_k \in FG$. Note, for example, that $C_1 = |G|^{-1}\rho_G$.

Clearly $C_1, \ldots, C_k \in \mathrm{cf}(G)$. In fact it is not difficult to see that the set $\{C_1, \ldots, C_k\}$ is a basis for the center of FG.

In the present context, however, the important point is that

$$C_r C_s = \sum_{t=1}^{k} n_{rst} C_t$$

for all r and s. Thus the calculation of the n_{rst} can be carried out via multiplications in the group algebra.

7.2 Dixon

Even for small examples such as those in the exercise above it soon becomes clear that Burnside's algorithm requires a considerable amount of calculation, particularly for step 3.

There are of course readily available computer routines for finding eigenvalues and eigenvectors. They tend, however, to do the calculations numerically rather than symbolically. As J. Dixon pointed out in [21], most theoretical uses of character tables require knowing the entries in exact form as algebraic integers. Consequently Dixon produced a revised form of the algorithm that overcame that difficulty, yet could be programmed for machine computation.

Some preparation is required.

Let $m = \exp(G)$, the least common multiple of the orders of elements of G. If $x \in G$ and $\chi_i \in \mathrm{Irr}(G)$ then $\chi_i(x)$ is a sum of complex mth roots of unity. If $\zeta \in \mathbb{C}$ is a primitive mth root of unity then all the character values $\chi_i(x)$ lie in $\mathbb{Z}[\zeta] = \{f(\zeta) : f(x) \in \mathbb{Z}[x]\}$.

Choose a prime $p \in \mathbb{Z}$ such that $p > 2\sqrt{|G|}$ and $m \mid p - 1$. Thus p is in the arithmetic progression $\{km + 1 : k \in \mathbb{N}\}$, and the existence of such a prime is the result of a deep theorem of Dirichlet (e.g. see [11], page 120). It is interesting to note that the usual proof of Dirichlet's theorem utilizes the theory of characters of abelian groups.

Since all prime divisors of $|G|$ are divisors of m it is clear that $p \nmid |G|$. The requirement that $p > 2\sqrt{|G|}$ ensures that $p > \chi_i(1)$ for all $\chi_i \in \mathrm{Irr}(G)$ by Theorem 5.2.14.

Next choose $z \in \mathbb{Z}$ having multiplicative order m when viewed as an element of $\mathbb{Z}_p^*$. Then z is a root mod p of the cyclotomic polynomial $\Phi_m(x) \in \mathbb{Z}[x]$. Since $\Phi_m(x)$ is the minimal polynomial for ζ over $\mathbb{Q}$ it follows that $\Phi_m(x) \mid f(x)$ for any $f(x) \in \mathbb{Z}[x]$ satisfying $f(\zeta) = 0$. As a consequence there is a well-defined map $\theta : \mathbb{Z}[\zeta] \to \mathbb{Z}_p$ given by $\theta : f(\zeta) \mapsto f(z) \pmod{p}$. Clearly θ is a ring homomorphism; it maps $\mathbb{Z}[\zeta]$ onto $\mathbb{Z}_p$ since $\theta(1) = 1 \pmod{p}$ is in its image. Extend θ in the obvious fashion so that it maps vectors with entries from $\mathbb{Z}[\zeta]$ to vectors with entries from $\mathbb{Z}_p$.

We continue using the notation of the previous section.

Proposition 7.2.1 *The image under the map θ of $\{v_1, \ldots, v_k\}$ is a set of k linearly independent eigenvectors of the matrices $N_1, \ldots, N_k$ (mod p); the eigenvalue corresponding to $\theta(v_i)$ for N_r is $\theta(\omega_{ir})$.*

Proof Only the linear independence needs proving – the rest is a consequence of the fact that θ is a ring homomorphism.

Define a weighted Hermitian inner product on the space of k-dimensional column vectors over $\mathbb{C}$ as follows: if $u = (u_1, \ldots, u_k)^t$ and $w = (w_i, \ldots, w_k)^t$ set

$$\langle u, w \rangle = |G|^{-1} \sum_i \frac{u_i \overline{w_i}}{|K_i|} .$$

Then we may interpret Proposition 7.1.2 as saying

$$\langle v_i, v_j \rangle = \frac{\delta_{ij}}{\chi_i(1)\chi_j(1)},$$

and $\{v_1, \ldots, v_k\}$ is an orthogonal set relative to the new inner product.

Let $W = \mathbb{Z}_p^k$, the $\mathbb{Z}_p$-space of column vectors, and use θ to define a bilinear form on W, also denoted $\langle *, * \rangle$, via $\langle \theta u, \theta v \rangle = \theta \langle u, v \rangle$ for vectors $u, v \in \mathbb{Z}[\zeta]^k$ (this is well-defined since $p \nmid |G|$ and $p \nmid |K_i|$, all i, the denominators of the weights).

Observe now that

$$\langle \theta v_i, \theta v_j \rangle = \theta \left(\frac{\delta_{ij}}{\chi_i(1)\chi_j(1)} \right) = \frac{\delta_{ij}}{\chi_i(1)\chi_j(1)} \pmod{p}$$

(recall that p does not divide the character degrees). Thus $\theta v_1, \ldots, \theta v_k$ are orthogonal relative to the bilinear form and hence are linearly independent by the standard argument from linear algebra. $\triangle$

Dixon's idea was to use the homomorphism θ to transfer the calculations to $\mathbb{Z}_p$. One further technicality is required to enable us to recover the character values in F after the calculations mod p.

Suppose that T is an F-representation of G, with character $\chi \in \mathrm{Irr}(G)$ and $d = \chi(1)$. Suppose that $x \in G$ has order k (so $k \mid m$). Let $\xi = \zeta^{m/k}$, a primitive kth root of unity. Then $\chi(x)$ is the sum of the d eigenvalues of $T(x)$, each a kth root of unity in $\mathbb{C}$, so $\chi(x) = \xi^{s_1} + \xi^{s_2} + \cdots + \xi^{s_d}$, say. Note also that $\chi(x^\ell) = \xi^{s_1 \ell} + \xi^{s_2 \ell} + \cdots + \xi^{s_d \ell}$ for any $\ell \in \mathbb{N}$. Retain this notation in the statement of the next proposition.

Proposition 7.2.2 *For each $s \in \mathbb{N}$, $0 \leq s \leq k - 1$, denote by $\mu(s)$ the multiplicity of ξ^s as an eigenvalue of $T(x)$. Then*

$$\mu(s) = \frac{1}{k} \sum_{\ell=0}^{k-1} \chi(x^\ell)\xi^{-s\ell} .$$

Proof Recall that if $\alpha \in \mathbb{C}$ is an rth root of unity then

$$\alpha^0 + \alpha^1 + \cdots + \alpha^{r-1} = \begin{cases} r & \text{if } \alpha = 1, \\ 0 & \text{otherwise.} \end{cases}$$

We calculate:

$$
\begin{aligned}
\frac{1}{k}\sum_{\ell=0}^{k-1} \chi(x^\ell)\xi^{-s\ell} &= \frac{1}{k}\sum_{\ell}(\xi^{s_1\ell} + \cdots + \xi^{s_d\ell})\xi^{-s\ell} \\
&= \frac{1}{k}\sum_{\ell}\xi^{(s_1-s)\ell} + \cdots + \frac{1}{k}\sum_{\ell}\xi^{(s_d-s)\ell}.
\end{aligned}
$$

The ith summation adds to 0 if $s_i \neq s$, to k if $s_i = s$, and the proposition follows.   △

Corollary 7.2.3 *The multiplicity $\mu(s)$ is uniquely determined by*

$$\mu(s) = \frac{1}{k}\sum_{\ell=0}^{k-1} \theta(\chi(x^\ell))z^{-s\ell m/k} \ (mod\ p)$$

and the fact that $p > \mu(s) \in \mathbb{N}$.

Proof Since $\mu(s) \leq d$ this is a consequence of $p > 2\sqrt{|G|}$.   △

Corollary 7.2.4 $\chi(x) = \sum_{s=0}^{k-1} \mu(s)\xi^s$.

We may now state Dixon's variation of the Burnside algorithm.

The Burnside-Dixon Algorithm

1. Determine the conjugacy classes of G.

2. Calculate all n_{rst}, hence all matrices N_r.

3. Find k linearly independent eigenvectors $\theta v_1, \ldots, \theta v_k$ for the matrices $N_1, \ldots, N_k$ (mod p), normalized to have first entry 1. The jth entry of θv_i is $\theta \omega_{ij}$.

4. Calculate $\chi_i(1)$, all i, via $\langle \theta v_i, \theta v_i \rangle = 1/\chi_i(1)^2$ (mod p). There is a unique solution for $\chi_i(1) \in \mathbb{N}$ satisfying $\chi_i(1) < p/2$, as must be the case since $p > 2\sqrt{|G|}$.

5. Calculate $\theta(\chi_i(K_j)) = \theta(\omega_{ij})\chi_i(1)/|K_j|$, all i and j.

6. Calculate all $\chi_i(K_j)$ from $\theta(\chi_i(K_j))$ by means of the corollaries to Proposition 7.2.2.

To carry out step 3 for a given N_j the elements of $\mathbb{Z}_p$ can be tried in succession to see whether $N_j - \lambda I$, $\lambda \in \mathbb{Z}_p$, has a nontrivial nullspace – if so a basis for the resulting eigenspace is determined. Thus $V = \mathbb{Z}_p^k$ is split as a direct sum of eigenspaces, say $V = V_1 \oplus \cdots \oplus V_r$. The next N, say N_{j+1}, then acts by restriction on each of the V_i's of dimension more than 1 to see whether it will split them into sums of lower-dimensional subspaces. That continues until all the eigenspaces are of dimension 1.

For an example let us take $G = \mathrm{Aff}(F)$, for $F = \mathbb{F}_5$, so G is Frobenius of order $5 \cdot 4 = 20$. It has the presentation

$$G = \langle a, b \mid a^5 = b^4 = 1, \, ba = a^2 b \rangle.$$

It is easy to check that the conjugacy classes are

$$K_1 = \{1\}, \quad K_2 = \mathrm{cl}(a) = \{a^i : 1 \le i \le 4\}, \quad K_3 = \mathrm{cl}(b) = \{a^i b : 0 \le i \le 4\},$$

$$K_4 = \mathrm{cl}(b^2) = \{a^i b^2 : 0 \le i \le 4\}, \quad K_5 = \mathrm{cl}(b^3) = \{a^i b^3 : 0 \le i \le 4\}.$$

The matrices N_i are easily computed, e.g. via calculating in the group algebra. Since not all N_i will be needed for the algorithm we list only the two that will be used:

$$N_2 = \begin{bmatrix} 0 & 1 & 0 & 0 & 0 \\ 4 & 3 & 0 & 0 & 0 \\ 0 & 0 & 4 & 0 & 0 \\ 0 & 0 & 0 & 4 & 0 \\ 0 & 0 & 0 & 0 & 4 \end{bmatrix} \quad \text{and} \quad N_3 = \begin{bmatrix} 0 & 0 & 1 & 0 & 0 \\ 0 & 0 & 4 & 0 & 0 \\ 0 & 0 & 0 & 5 & 0 \\ 0 & 0 & 0 & 0 & 5 \\ 5 & 5 & 0 & 0 & 0 \end{bmatrix}.$$

We have $2\sqrt{|G|} = 4\sqrt{5} < 12$ and $\exp(G) = 20$, so we need a prime that is at least 12 and is $\equiv 1 \pmod{20}$. The first one is $p = 41$, so we calculate mod 41.

The only eigenvalues of N_2 in $\mathbb{Z}_{41}$ are 4 and $40 = -1$. The eigenspace for -1 is 1-dimensional; we label it V_5 and take as its normalized generator $\theta v_5 = (1, -1, 0, 0, 0)^t$. The eigenspace for 4 is 4-dimensional – the basis that results from Gaussian elimination is

$$\{(1, 4, 0, 0, 0), \, (0, 0, 1, 0, 0), \, (0, 0, 0, 1, 0), \, (0, 0, 0, 0, 1)\}.$$

If we denote the eigenspace by V_1 then we have $V = \mathbb{Z}_{41}^5 = V_1 \oplus V_5$. Next N_3 acts on V_1; its action there is represented by the 4×4 matrix

$$A = \begin{bmatrix} 0 & 1 & 0 & 0 \\ 0 & 0 & 5 & 0 \\ 0 & 0 & 0 & 5 \\ 5 & 0 & 0 & 0 \end{bmatrix}.$$

The matrix A has 4 distinct eigenvalues mod 41, viz. $\lambda = 0, 8, 9$, and $40 = -1$. For example, $\lambda = 9$ has as an eigenvector $(1, 5, 5, 5)^t$ and, using its entries as coefficients of the basis vectors for V_1, we obtain $\theta v_1 = (1,4,5,5,5)^t$. In like manner we obtain $\theta v_2 = (1,4,-4,-5,4)^t$, $\theta v_3 = (1,4,4,-5,-4)^t$, and $\theta v_4 = (1,4,-5,5,-5)^t$, and V_1 is split by N_3 into the direct sum of four 1-dimensional subspaces with the indicated normalized basis vectors.

As a sample of the determination of degrees note that

$$\langle \theta v_5, \theta v_5 \rangle = \frac{1}{20}\left[\frac{1 \cdot 1}{1} + \frac{(-1) \cdot (-1)}{4}\right] = \frac{1}{16} = \frac{1}{\chi_5(1)^2} \pmod{41},$$

so $\chi_5(1) = 4$. We see similarly that $\chi_1, \ldots, \chi_4$ are all linear.

Take $\zeta = e^{2\pi i/20}$, a primitive 20th root of unity in $\mathbb{C}$, and set $z = 2$, which has multiplicative order 20 in $\mathbb{Z}_{41}$. The final step is to calculate the other character values, using the fact that $\theta \chi_i(K_j) = (\chi_i(1)/|K_j|)\theta \omega_{ij}$. For example, let us evaluate $\chi_5(K_2) = \chi_5(a)$. Since $|a| = 5$ we use $\xi = \zeta^4$ and correspondingly use $z^4 = 2^4$ in $\mathbb{Z}_{41}$.

We need the multiplicities $\mu(s)$:

$$\mu(0) = \frac{1}{5}\left[4 - \frac{4}{4} - 1 - 1 - 1\right] = 0,$$

$$\mu(1) = \frac{1}{5}\left[4 - \left(\frac{4}{4}\right) \cdot 2^{-4} - 1 \cdot 2^{-8} - 1 \cdot 2^{-12} - 1 \cdot 2^{-16}\right] = 1$$

(calculating in $\mathbb{Z}_{41}$), and similarly $\mu(2) = \mu(3) = \mu(4) = 1$. Thus

$$\chi_5(K_2) = \xi + \xi^2 + \xi^3 + \xi^4 = -1,$$

since $\xi = \zeta^4$ is a primitive 5th root of unity. The complete character table follows.

	K_1	K_2	K_3	K_4	K_5
χ_1	1	1	1	1	1
χ_2	1	1	-1	1	-1
χ_3	1	1	$-i$	-1	i
χ_4	1	1	i	-1	$-i$
χ_5	4	-1	0	0	0

7.3 Schneider

Computer implementations of the Burnside-Dixon algorithm were written by several people, beginning with Dixon himself. In 1990 G. Schneider discovered a variation on the algorithm that in many cases speeds it up considerably. The variation will be described briefly below, after some further preliminary discussion.

We retain the notation of the previous two sections. Recall, in particular, that $K_{j'} = K_j^{-1}$ for each conjugacy class K_j. We require first some elementary information concerning the entries n_{rst} of the class matrix N_r.

Proposition 7.3.1 *For all values of r, s, and t we have*

1. $n_{rst} = n_{s'r't'}$,

2. $n_{rst} = n_{srt}$, *and*

3. $n_{rst}|K_t| = n_{ts'r}|K_r|$.

Proof For parts 1 and 2 note that $xy = z$ if and only if $y^{-1}x^{-1} = z^{-1}$ if and only if $yx = z^x$. For part 3 write $K_r = \{x_1, \ldots, x_u\}$ and $K_s = \{y_1, \ldots, y_v\}$, say. Note that $|x_iK_s \cap K_t| = |x_jK_s \cap K_t|$ for all i, j, and the sets are disjoint if $i \neq j$. Since $(x_1K_s \cap K_t) \cup \cdots \cup (x_uK_s \cap K_t)$ is the set of all elements of K_t exhibited as products x_iy_j in all possible ways we see, writing $x_1 = x$, that $|K_r||xK_s \cap K_t| = n_{rst}|K_t|$. We may assume the y_i's labeled so that $xK_s \cap K_t = \{xy_1, \ldots, xy_w\}$, and set $xy_i = z_i$, $1 \leq i \leq w$. Thus $x = z_1y_1^{-1} = \cdots = z_wy_w^{-1}$ are precisely the distinct ways that $x \in K_r$ can be written as a product of elements from K_t and $K_{s'}$, i.e. $w = n_{ts'r}$, and the result follows. $\triangle$

Exercise

Show that $|K_r||K_s| = \sum_t n_{rst}|K_t|$.

Proposition 7.3.2 *If $1 \leq r, s \leq k$ then $\sum_{t=1}^{k} \chi_i(K_t)n_{str} = \omega_{is'}\chi_i(K_r)$.*

Proof By Proposition 5.2.16 we have

$$\omega_{ir}\omega_{is'} = \sum_t n_{rs't}\omega_{it},$$

which can be written as

$$\omega_{is'} \cdot \frac{|K_r|\chi_i(K_r)}{\chi_i(1)} = \sum_t n_{rs't} \cdot \frac{|K_t|\chi_i(K_t)}{\chi_i(1)}.$$

Cancel the common factor of $1/\chi_i(1)$, and apply part 3 of Proposition 7.3.1. The equation becomes

$$\omega_{is'}|K_r|\chi_i(K_r) = \sum_t n_{tsr}|K_r|\chi_i(K_t).$$

Apply part 2 of Proposition 7.3.1 and cancel the common $|K_r|$ to complete the proof. $\triangle$

Proposition 7.3.2 admits the following interpretation: the row vector $v = (\chi_i(K_1), \ldots, \chi_i(K_k))$, which is just row i of the character table, is a *left* eigenvector of the class matrix N_s with eigenvalue $\omega_{is'}$, i.e.XS $vN_s = \omega_{is'}v$.

The chief difference in Schneider's version of the algorithm is to use Proposition 7.3.2, hence to calculate left eigenspaces rather than right eigenspaces, although it should be pointed out that he introduced further refinements as well. In particular, he first calculates all the linear characters as characters of G/G'. He incorporates tests that allow for calculating only certain columns of some of the N_s, and for avoiding calculation of N_s if it will not contribute to splitting of eigenspaces. Furthermore he has developed an independent method that sometimes works to complete the splitting of 2-dimensional eigenspaces. See [56] for details.

The calculations are done in $\mathbb{Z}_p$ for an appropriate prime p, and the actual values in $\mathbb{C}$ are recovered just as in Dixon's version.

The Dixon-Schneider version of the algorithm is incorporated into the system GAP (see [57]). For example, GAP produced the following character table of the Mathieu group M_{11} (see Chapter 2).

	K_1	K_2	K_3	K_4	K_5	K_6	K_7	K_8	K_9	K_{10}
χ_1	1	1	1	1	1	1	1	1	1	1
χ_2	10	2	2	0	0	0	-1	1	-1	-1
χ_3	10	0	-2	α	$-\alpha$	0	1	1	-1	-1
χ_4	10	0	-2	$-\alpha$	α	0	1	1	-1	-1
χ_5	11	-1	3	-1	-1	1	0	2	0	0
χ_6	16	0	0	0	0	1	0	-2	β	γ
χ_7	16	0	0	0	0	1	0	-2	γ	β
χ_8	44	0	4	0	0	-1	1	-1	0	0
χ_9	45	1	-3	-1	-1	0	0	0	1	1
χ_{10}	55	-1	-1	1	1	0	-1	1	0	0

Character Table of M_{11}

Here $\alpha = \zeta + \zeta^3$, $\beta = \xi^2 + \xi^6 + \xi^7 + \xi^8 + \xi^{10}$, and $\gamma = \xi + \xi^3 + \xi^4 + \xi^5 + \xi^9$, with $\zeta \in \mathbb{C}$ a primitive 8th root of unity and $\xi \in \mathbb{C}$ a primitive 11th root of unity.

The actual GAP commands used to create the table are as follows.

```
gap> m11 := Group((1,2,3,4,5,6,7,8,9,10,11),
(3,7,11,8)(4,10,5,6));
gap> m11.name := ''m11'';
gap> CharTable(m11);
```

Exercise

Use GAP to find the character table of the Mathieu group M_{12}.

Chapter 8

Characters of S_n and A_n

There is little doubt that the world's best-loved finite (nonabelian) groups are the symmetric groups S_n and the alternating groups A_n. In this chapter we present reasonably practical algorithms for calculating their character tables.

8.1 Symmetric Groups

If $0 < n \in \mathbb{N}$ recall that a *partition* of n is a sequence $\lambda = (\lambda_1, \ldots, \lambda_r)$ of positive integers (the *parts* of λ), satisfying

$$\lambda_1 \geq \lambda_2 \geq \cdots \geq \lambda_r$$

and

$$\lambda_1 + \lambda_2 + \cdots + \lambda_r = n.$$

It will be convenient to use another form of exponential notation for partitions – one example should make it clear. The partition $\lambda = (4, 4, 3, 2, 2, 2, 1)$ of $n = 18$ will be written as $\lambda = (4^2, 3, 2^3, 1)$.

The partitions of n determine the conjugacy classes in S_n – if K is a conjugacy class and $\sigma \in K$ write $\sigma = \sigma_1 \sigma_2 \cdots \sigma_r$, a product of disjoint cycles (including 1-cycles), with $|\sigma_i| \geq |\sigma_j|$ for $i > j$. Setting $\lambda_i = |\sigma_i|$, all i, we have a partition $\lambda = (\lambda_1, \ldots, \lambda_r)$ of n. It will be convenient to refer to σ as an element of *type* λ (note that this is related to, but not the same as, the usual notion of *cycle type* of σ).

It is often convenient to agree that each partition λ of n has n parts, which can be attained by simply adjoining sufficiently many 0 parts at the end.

Each partition λ of n has a *Young diagram*, which can be thought of as a (left-aligned) stack of empty boxes, the number of boxes in each row being the parts of λ. For example, the partition $\lambda = (3, 2^2, 1^2)$ of $n = 9$ has the

161

Young diagram

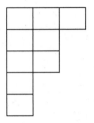

Each Young diagram has an *opposite* diagram (sometimes called the *conjugate* diagram), obtained by transposing it. If the diagram is of type λ we will denote the type of the opposite diagram by λ'. For example if $\lambda = (3, 2^2, 1^2)$, as above, then $\lambda' = (5, 3, 1)$, with Young diagram

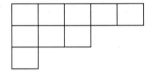

If $\lambda = (\lambda_1, \lambda_2, \ldots)$ and $\lambda' = (\lambda'_1, \lambda'_2, \ldots)$ then it is easy to see that

$$\lambda'_i = |\{\lambda_j \colon \lambda_j \geq i\}|.$$

Let us write $\boxed{n}$ for the set $\{1, 2, \ldots, n\}$. Thus, for example, $S_n = \mathrm{Perm}(\boxed{n})$. A *set partition* Λ of $\boxed{n}$ is as usual a partitioning of $\boxed{n}$ into a disjoint union of nonempty subsets, say $\boxed{n} = \Lambda_1 \dot\cup \cdots \dot\cup \Lambda_r$, labeled so that $|\Lambda_i| \geq |\Lambda_j|$ if $i > j$. If we set $\lambda_i = |\Lambda_i|$, all i, it is immediate that $\lambda = (\lambda_1, \ldots, \lambda_r)$ is a partition of n, which we will call the *type* of Λ; we write $\overline{\Lambda} = \lambda$.

The symmetric group S_n acts naturally on the family of all set partitions of $\boxed{n}$; the orbits are determined by types. If Λ is a set partition of type λ then the stabilizer of Λ in S_n will be denoted S_Λ and is called a *Young subgroup* of type Λ (or type λ). If the sets in the partition are $\Lambda_1, \ldots, \Lambda_r$ then $S_\Lambda = \mathrm{Perm}(\Lambda_1) \times \cdots \times \mathrm{Perm}(\Lambda_r)$.

If the elements of $\boxed{n}$ are distributed among the boxes of a Young diagram the resulting object is called a *Young tableau*. For $\lambda = (3, 2^2, 1^2)$, as above, one possible Young tableau is

3	1	6
7	9	
2	4	
5		
8		

The rows of a Young tableau determine a set partition of $\boxed{n}$, as do the columns. For the example above the row partition is

$$(\{3,1,6\}, \{7,9\}, \{2,4\}, \{5\}, \{8\}),$$

and the column partition is

$$(\{3,7,2,5,8\}, \{1,9,4\}, \{6\}).$$

The Young subgroup of the row partition is

$$\mathrm{Perm}(\{3,1,6\}) \times \mathrm{Perm}(\{7,9\}) \times \mathrm{Perm}(\{2,4\}) \times \mathrm{Perm}(\{5\}) \times \mathrm{Perm}(\{8\})$$

$$\cong S_3 \times S_2 \times S_2 \times S_1 \times S_1$$

and that of the column partition is

$$\mathrm{Perm}(\{3,7,2,5,8\}) \times \mathrm{Perm}(\{1,9,4\}) \times \mathrm{Perm}(\{6\})$$

$$\cong S_5 \times S_3 \times S_1.$$

Note that the row partition and column partition are of opposite types. It is clear that the intersection of any set in the row partition with any set in the column partition has at most 1 element in it. As a consequence their Young subgroups intersect trivially.

In general, two set partitions $\Lambda = (\Lambda_1, \cdots, \Lambda_r)$ and $M = (M_1, \cdots, M_s)$ of $\boxed{n}$ are said to be *disjoint* if $|\Lambda_i \cap M_j| \leq 1$ for all i and j. They are called *opposite* if they occur as the row partition and column partition for some Young tableau. Thus opposite set partitions are disjoint. Note that a given set partition typically has several opposites; *e.g.* $\Lambda = (\{1,2,3\}, \{4\})$ is opposite to each of $M_1 = (\{1,4\}, \{2\}, \{3\})$, $M_2 = (\{2,4\}, \{1\}, \{3\})$, and $M_3 = (\{3,4\}, \{2\}, \{1\})$.

Proposition 8.1.1 *If Λ is a set partition of $\boxed{n}$ then the Young subgroup S_Λ acts transitively on the collection of all partitions that are opposite to Λ.*

Proof If T_1 and T_2 are two Young tableaux both having Λ as row partition then T_1 can be transformed to T_2 by permuting the entries of its rows, i.e. by an element of S_Λ. $\triangle$

Proposition 8.1.2 *Set partitions Λ and M of $\boxed{n}$ are opposite if and only if they are disjoint and of opposite types.*

Proof $\Rightarrow$: This is clear.
$\Leftarrow$: There is a simple algorithm for producing a Young tableau T whose row partition is Λ and column partition is M: the ij-entry of T (when it exists) is the unique element of $\Lambda_i \cap M_j$. The proof that this works becomes clear (and is easily formalized) after seeing an example. See the following exercise. $\triangle$

Exercise

Apply the algorithm suggested in the proof above to the set partitions

$$\Lambda = (\{2,6,9\},\{4,7\},\{3,8\},\{1\},\{5\})$$

and

$$M = (\{1,3,5,7,9\},\{4,6,8\},\{2\}).$$

There is a useful partial ordering on the set of partitions of n. If $\lambda = (\lambda_1,\ldots,\lambda_n)$ and $\mu = (\mu_1,\ldots,\mu_n)$ say that λ is *dominated* by μ, and write $\lambda \ll \mu$ (or $\mu \gg \lambda$), if

$$\sum_{i=1}^{k} \lambda_i \leq \sum_{i=1}^{k} \mu_i$$

for $1 \leq k \leq n$. Thus, for example, $(2,2,2) \ll (3,2,1)$, but $(2,2,2)$ and $(3,1,1,1)$ are not comparable.

Define a lexicographic total ordering on the set of partitions of n as follows. If $\lambda = (\lambda_1,\ldots,\lambda_n)$ and $\mu = (\mu_1,\ldots,\mu_n)$ are partitions, write $\lambda \leq \mu$ if $\lambda = \mu$ or if for some k we have $\lambda_i = \mu_i$ for $i < k$ but $\lambda_k < \mu_k$.

Exercises

1. Show that dominance and the lexicographic ordering are the same if $n \leq 5$.

2. Show in general that if $\lambda \ll \mu$ then $\lambda \leq \mu$.

Proposition 8.1.3 *If Λ and M are disjoint set partitions of $\boxed{n}$ having types λ and μ, then $\lambda \ll \mu'$.*

Proof Let T_λ and T_μ be Young tableaux whose row partitions are Λ and M, respectively. Imagine distributing the entries of T_λ, row by row, into the boxes of the diagram of T_μ. To begin, the λ_1 entries of row 1 of T_λ must go into different rows of T_μ, of which there are μ'_1, so $\lambda_1 \leq \mu'_1$. Assume inductively that the entries of the first $k-1$ rows of T_λ have been distributed and that $\sum_{i=1}^{\ell} \lambda_i \leq \sum_{i=1}^{\ell} \mu'_i$, $1 \leq \ell \leq k-1$. Since row entries in T_μ can be rearranged without impairing disjointness we may assume that the entries of the first $k-1$ rows of T_λ have been distributed among the first $k-1$ columns of T_μ. Thus the number of still vacant boxes in the first k columns of T_μ is $\sum_{i=1}^{k} \mu'_i - \sum_{i=1}^{k-1} \lambda_i$, and there are at most that many rows of T_μ with vacant boxes. Since it is possible to distribute the entries of row k of T_λ into different rows of T_μ, it must be the case that $\lambda_k \leq \sum_{i=1}^{k} \mu'_i - \sum_{i=1}^{k-1} \lambda_i$, as required. $\triangle$

The converse of the proposition above is also true (but more difficult to prove); the two together are a version of the Gale-Ryser Theorem, an important result in combinatorics. For an amusing proof see [27].

Corollary 8.1.4 *Suppose that λ and μ are partitions of n. If there exist Young subgroups of types λ and μ having trivial intersection then $\lambda \ll \mu'$.*

Suppose now that λ is a partition of n. Choose a set partition Λ of type λ and get thereby a Young subgroup S_Λ. Set $\psi_\lambda = 1_{S_\Lambda}^{S_n}$, the principal character of S_Λ induced to S_n. Since ψ_λ is the permutation character of the action of S_n on the set of all set partitions of type λ it is independent of the particular choice of Λ.

Denote by ϵ_Λ the alternating character of S_Λ and set $\varphi_\lambda = \epsilon_\Lambda^{S_n}$, also independent of the choice of Λ.

Proposition 8.1.5 *If λ and μ are partitions of n and $(\psi_\lambda, \varphi_\mu) \neq 0$, then $\lambda \ll \mu'$.*

Proof Choose partitions Λ and M of $\boxed{n}$ of types λ and μ, and denote the respective Young subgroups by K and H. By the Mackey Theorem 6.4.2 we have

$$(\psi_\lambda, \varphi_\mu) = \sum_{i=1}^{s} (1_{H_i}, \epsilon_{H_i}^{x_i}),$$

where $H_i = K^{x_i} \cap H$, the x_i's being $(K\text{-}H)$–double coset representatives, with $x_1 = 1$. Thus each H_i is an intersection of Young subgroups of types λ and μ. Note that $\epsilon_{H_i}^{x_i} = \epsilon_{H_i}$, since parity is preserved by conjugation. If any $H_i \neq 1$ then it is a nontrivial Young subgroup; hence it contains an odd permutation, so $1_{H_i} \neq \epsilon_{H_i}$ and $(1_{H_i}, \epsilon_{H_i}) = 0$. Thus for $(\psi_\lambda, \varphi_\mu)$ to be nonzero there must exist disjoint partitions of types λ and μ, in which case $\lambda \ll \mu'$ by Proposition 8.1.3. $\triangle$

Theorem 8.1.6 *If λ is a partition of n then $(\psi_\lambda, \varphi_{\lambda'}) = 1$, so ψ_λ and $\varphi_{\lambda'}$ have a unique common irreducible constituent, occurring with multiplicity 1 in each of them; it will be denoted χ_λ.*

Proof Choose opposite partitions Λ and Λ' of $\boxed{n}$ of types λ and λ', and denote the respective Young subgroups by K and H, so $K \cap H = 1$. As in the proof of Proposition 8.1.5 we have contributions of 1 to $(\psi_\lambda, \varphi_{\lambda'})$ from each $H_i = 1$, and $H_1 = K \cap H = 1$, so it must be shown that $H_i \neq 1$ for $i > 1$. Suppose then that $H_i = K^{x_i} \cap H = 1$. Thus Λ^{x_i} and Λ' are disjoint, hence opposite by Proposition 8.1.2. By Proposition 8.1.1 we have $\Lambda^{x_i h} = \Lambda$ for some $h \in H$, so $x_i h \in K$ and $x_i \in KH$, the $(K\text{-}H)$–double coset containing $x_1 = 1$; hence $x_i = x_1$ and $i = 1$. $\triangle$

For each partition λ of n denote by K_λ the conjugacy class in S_n of elements of type λ. Order the classes according to the lexicographic ordering of the partitions. Thus $K_{(1^n)}$ is first and $K_{(n)}$ is last. Order the collection of characters χ_λ according to reverse (descending) lexicographical ordering of partitions, so $\chi_{(n)}$ is first and $\chi_{(1^n)}$ is last. Then form the square matrix $X = [\chi_\lambda(K_\mu)]$. With the same indexing of rows and columns define $Y = [\psi_\lambda(K_\mu)]$.

Proposition 8.1.7 *If χ_ν is a constituent of ψ_λ then $\lambda \leq \nu$.*

Proof Since χ_ν is also a constituent of $\varphi_{\nu'}$ we have $(\psi_\lambda, \varphi_{\nu'}) \neq 0$, so $\lambda \ll \nu$ by Proposition 8.1.3. Thus $\lambda \leq \nu$ by Exercise 2 on page 164. $\triangle$

Theorem 8.1.8 *The characters χ_λ are all distinct, so X is the character table of S_n.*

Proof Suppose that $\chi_\lambda = \chi_\mu$ for partitions λ and μ. Then χ_λ is a constituent of ψ_μ and χ_μ is a constituent of ψ_λ, so $\lambda \leq \mu \leq \lambda$ by Proposition 8.1.7, and $\lambda = \mu$. $\triangle$

Corollary 8.1.9 $\psi_\lambda = \sum\{(\psi_\lambda, \chi_\nu)\chi_\nu : \nu \geq \lambda\}$.

If λ and μ are partitions of n then a μ-*refinement* of λ is an expression of each of the parts of λ as a sum of parts of μ, the parts of μ being exhausted in the process. An example should make this clear. Take $\lambda = (5, 4)$ and $\mu = (3, 2^2, 1^2)$, which we write as $(3, 2_1, 2_2, 1_1, 1_2)$. Then there are 5 different μ-refinements of λ:

$$(3 + 2_1, 2_2 + 1_1 + 1_2), \quad (3 + 2_2, 2_1 + 1_1 + 1_2), \quad (3 + 1_1 + 1_2, 2_1 + 2_2),$$

$$(2_1 + 2_2 + 1_1, 3 + 1_2), \quad \text{and} \quad (2_1 + 2_2 + 1_2, 3 + 1_1).$$

Proposition 8.1.10 *If λ and μ are partitions of n then the character value $\psi_\lambda(K_\mu)$ is the number of distinct μ-refinements of λ.*

Proof Recall that ψ_λ is the permutation character of S_n on the collection of all set partitions of $\boxed{n}$ of type λ. Choose $\sigma \in K_\mu$ – we need to count the set partitions of type λ that are fixed by σ. Write $\sigma = \sigma_1 \cdots \sigma_s$, with σ_i a μ_i-cycle, all i, and take $M = (M_1, \ldots, M_s)$, the corresponding set partition, i.e. M_i is the orbit of σ_i. If $\Lambda = (\Lambda_1, \ldots, \Lambda_r)$ is of type λ, then σ fixes Λ if and only if each M_j is contained in some Λ_i, which is clearly if and only if M determines a μ-refinement of λ. $\triangle$

Thus calculation of the matrix Y is a relatively straightforward combinatorial problem.

Proposition 8.1.11 *If λ is a partition of n then $\chi_{\lambda'} = \epsilon \cdot \chi_\lambda$, where as usual ϵ denotes the alternating character of S_n.*

Proof We have $\varphi_\lambda = \epsilon_{S_\lambda}^{S_n}$ and $\epsilon \cdot \psi_\lambda = \epsilon \cdot 1_{S_\lambda}^{S_n} = (\epsilon|_{S_\lambda} \cdot 1_{S_\lambda})^{S_n} = \epsilon_{S_\lambda}^{S_n}$ by Proposition 6.2.7, so $\varphi_\lambda = \epsilon \cdot \psi_\lambda$. Now χ_λ is the unique common constituent of ψ_λ and $\varphi_{\lambda'}$, so $\epsilon\chi_\lambda$ is the unique common constituent of $\epsilon\psi_\lambda = \varphi_\lambda$ and $\epsilon\varphi_{\lambda'} = \psi_{\lambda'}$, which is $\chi_{\lambda'}$. Thus $\chi_{\lambda'} = \epsilon\chi_\lambda$. $\triangle$

The results above make possible an efficient recursive algorithm for calculation of the character table X, assuming that Y has been calculated. In fact because of Proposition 8.1.11 it is only necessary to calculate approximately half of the rows of X, and as a consequence only those rows of Y are needed; denote the truncated versions by X' and Y'.

Since $\chi_{(n)} = 1_{S_n}$ the first row of X' has all entries 1. If χ_ν has been calculated for all $\nu > \lambda$ then by Corollary 8.1.9 we see that $\chi_\lambda = \psi_\lambda - \sum\{(\psi_\lambda, \chi_\nu)\chi_\nu : \nu > \lambda\}$. Thus we need only take inner products of row λ of Y' with the already constructed rows of X', subtract those multiples of the rows of X' from row λ of Y', and insert the result as row λ of X'.

To illustrate, let us calculate the character table of S_5. We will simply index the rows and columns of both X and Y' by the partitions of 5, ordered as indicated above. The sizes of the classes have been inserted in Y' to facilitate taking inner products.

First Y'.

	(1^5)	$(2,1^3)$	$(2^2,1)$	$(3,1^2)$	$(3,2)$	$(4,1)$	(5)
$\|K_\lambda\|$	1	10	15	20	20	30	24
(5)	1	1	1	1	1	1	1
$(4,1)$	5	3	1	2	0	1	0
$(3,2)$	10	4	2	1	1	0	0
$(3,1^2)$	20	6	0	2	0	0	0

Then we begin with $\chi_{(5)} = 1_{S_5}$. Since $(\psi_{(4,1)}, \chi_{(5)}) = 1$ we have

$$\chi_{(4,1)} = \psi_{(4,1)} - \chi_{(5)}.$$

Next $(\psi_{(3,2)}, \chi_{(5)}) = (\psi_{(3,2)}, \chi_{(4,1)}) = 1$, so

$$\chi_{(3,2)} = \psi_{(3,2)} - \chi_{(5)} - \chi_{(4,1)}.$$

Finally $(\psi_{(3,1^2)}, \chi_{(5)}) = 1$, $(\psi_{(3,1^2)}, \chi_{(4,1)}) = 2$, and $(\psi_{(3,1^2)}, \chi_{(3,2)}) = 1$, so

$$\chi_{(3,1^2)} = \psi_{(3,1^2)} - \chi_{(5)} - 2\chi_{(4,1)}) - \chi_{(3,2)}.$$

The remaining rows of X are then obtained via Proposition 8.1.11. Note that $(3,1^2)' = (3,1^2)$, so $\chi_{(3,1^2)} = \epsilon \cdot \chi_{(3,1^2)}$.

	(1^5)	$(2,1^3)$	$(2^2,1)$	$(3,1^2)$	$(3,2)$	$(4,1)$	(5)
(5)	1	1	1	1	1	1	1
$(4,1)$	4	2	0	1	-1	0	-1
$(3,2)$	5	1	1	-1	1	-1	0
$(3,1^2)$	6	0	-2	0	0	0	1
$(2^2,1)$	5	-1	1	-1	-1	1	0
$(2,1^3)$	4	-2	0	1	1	0	-1
(1^5)	1	-1	1	1	-1	-1	1

Characters of S_5

Exercise

Determine the character tables of S_6 and S_7.

8.2   Alternating Groups

Let us assume throughout this section that $n \geq 2$, so $[S_n : A_n] = 2$.

Since A_n is a normal subgroup of S_n it is a union of S_n-conjugacy classes. If K is an S_n-conjugacy class and $K \subseteq A_n$ there are two possibilities: K may also be an A_n-conjugacy class, or it may split up as a union of smaller A_n-classes.

Choose $\sigma \in K$. Observe that $C_{A_n}(\sigma) = A_n \cap C_{S_n}(\sigma)$, so that $C_{A_n}(\sigma)$ has index either 1 or 2 in $C_{S_n}(\sigma)$. If the index is 2 then $|K| = |S_n|/|C_{S_n}(\sigma)| = (2|A_n|)/(2|C_{A_n}(\sigma)|) = |\mathrm{cl}_{A_n}(\sigma)|$, so $K = \mathrm{cl}_{A_n}(\sigma)$. If the index is 1 (i.e. $C_{S_n}(\sigma) \leq A_n$) then $|K| = [S_n : C_{S_n}(\sigma)] = [S_n : C_{A_n}(\sigma)] = 2[A_n : C_{A_n}(\sigma)] = 2|\mathrm{cl}_{A_n}(\sigma)|$. Thus K splits in that case into two A_n-classes, each of size $|K|/2$.

We see then that the key issue is whether or not there is an odd element of S_n that commutes with σ.

Proposition 8.2.1 *If K is a conjugacy class in S_n, $K \subseteq A_n$, and $\sigma \in K$, then K is a conjugacy class in A_n if and only if some odd element of S_n commutes with σ; if that is not the case then K splits as the union of two A_n-classes, each of size $|K|/2$. If λ is the (partition) type of σ then K splits if and only if the parts of λ are all odd and all different from each other.*

Proof   The first statement is clear from the discussion above.

If λ has an even part then σ has a cycle of that length, which is odd and commutes with σ. If λ has 2 odd parts of the same length it is also possible to construct an odd element that commutes with σ – an example will make this clear. If σ has $(abc)(def)$ in its cycle decomposition then $\tau = (ad)(be)(cf)$ is odd, and τ commutes with σ since conjugation of σ by τ simply permutes the cycles (abc) and (def). On the other hand, if the parts of λ are odd and distinct, and if some $\tau \in S_n$ commutes with σ, then inspection of $\sigma^\tau = \sigma$ shows that the cycles of τ are either 1-cycles or have the same (odd) lengths as cycles of σ; hence τ must be even.   △

In the case of splitting we may clearly take any $\sigma \in K$ as a representative of one A_n-class, then σ^τ as a representative of the other, where τ is any odd element in S_n, e.g. $\tau = (12)$.

For an example, note that the even classes in S_6 are those of types (1^6), $(2^2, 1^2)$, $(3, 1^3)$, (3^2), $(4, 2)$, and $(5, 1)$. Of these only the class of 144 5-cycles, of type $(5, 1)$, splits, and as representatives of the 2 A_n-classes we may take $\sigma = (12345)$ and $\sigma^\tau = (12345)^{(12)} = (21345)$.

Now the characters.

If $\lambda \neq \lambda'$ then it follows from Proposition 8.1.11 that χ_λ and $\chi_{\lambda'}$ agree on the classes that lie in A_n. By Proposition 6.4.5 those characters restrict to irreducible characters of A_n.

If $\lambda = \lambda'$ then χ_λ splits as a sum of 2 conjugate characters, also by Proposition 6.4.5, say $\chi_\lambda|_{A_n} = \chi_\lambda^{(1)} + \chi_\lambda^{(2)}$, each of degree $\chi_\lambda(1)/2$.

Since $\lambda = (\lambda_1, \ldots, \lambda_r) = \lambda'$ the Young diagram of λ is symmetric; it can be thought of as made up of "hooks", the first hook being the first row together with the first column. Let us write s for the total number of hooks, and δ_i for the length of the ith hook, i.e. the number of boxes in the hook. Thus $\delta_1 = 2\lambda_1 - 1$, $\delta_2 = 2(\lambda_2 - 1) - 1 = 2\lambda_2 - 3$, and so on.

For example, $\lambda = (4, 3, 2, 1)$ has diagram

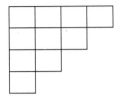

with two hooks of lengths $\delta_1 = 7$ and $\delta_2 = 3$.

The hook lengths of λ determine a partition $\delta = (\delta_1, \ldots, \delta_s)$; it is clear from Proposition 8.2.1 that K_δ splits in A_n into two classes, call them $K_\delta^{(1)}$ and $K_\delta^{(2)}$.

For each class K_μ, with $\mu \neq \delta$, we have $\chi_\lambda^{(i)}(K_\mu) = \chi_\lambda(K_\mu)/2$, and it remains only to determine the entries in the 2×2 box where the two split characters $\chi_\lambda^{(i)}$ take their values on the two split classes $K_\delta^{(i)}$.

Set $\Delta = \prod_{i=1}^s \delta_i$.

To complete the story, note first that $\chi_\lambda(K_\delta) = (-1)^{(n-s)/2}$; abbreviate it as ι. Then we may write

$$\chi_\lambda^{(1)}(K_\delta^{(1)}) = \frac{\iota + \sqrt{\iota \cdot \Delta}}{2}, \quad \chi_\lambda^{(1)}(K_\delta^{(2)}) = \frac{\iota - \sqrt{\iota \cdot \Delta}}{2}$$

and

$$\chi_\lambda^{(2)}(K_\delta^{(1)}) = \frac{\iota - \sqrt{\iota \cdot \Delta}}{2}, \quad \chi_\lambda^{(2)}(K_\delta^{(2)}) = \frac{\iota + \sqrt{\iota \cdot \Delta}}{2}.$$

The proof of this is surprisingly(?) long and involved and will not be repeated here. See [5] or [40].

For an example we may determine the character table of A_5 from the table for S_5 in the previous section. The even classes are (1^5), $(2^2, 1)$, $(3, 1^2)$, and (5), and only (5) splits into two A_5-classes, which we will denote by (5_1) and (5_2). The first three characters of S_5 restrict to irreducible characters of A_5; denote them simply by χ_1, χ_2, and χ_3. Only $\chi_{(3,1^2)}$ splits, and $\lambda = (3, 1^2)$ has hook length partition $\delta = (5)$, since λ has only one hook. For that pair λ, δ we have $\iota = (-1)^{(5-1)/2} = 1$ and $\Delta = 5$. Denote the two conjugate characters that are constituents of $\chi_{(3,1^2)}|A_5$ by χ_4 and χ_5, and we have the character table as follows.

	(1^5)	$(2^2,1)$	$(3,1^2)$	(5_1)	(5_2)
χ_1	1	1	1	1	1
χ_2	4	0	1	-1	-1
χ_3	5	1	-1	0	0
χ_4	3	-1	0	$\frac{1+\sqrt{5}}{2}$	$\frac{1-\sqrt{5}}{2}$
χ_5	3	-1	0	$\frac{1-\sqrt{5}}{2}$	$\frac{1+\sqrt{5}}{2}$

Characters of A_5

Compare the table with the one on page 134.

Exercise

Determine the character tables of A_6 and A_7.

The character tables of the alternating groups A_n, for $5 \leq n \leq 13$, appear explicitly in the Atlas of Finite Groups [14], and a scheme appears in the Atlas for constructing the character tables of the corresponding symmetric groups.

There is a formidable literature on the representation theory of S_n (and to a lesser extent that of A_n), including a great variety of combinatorial algorithms and the general theory of symmetric functions. Two recent references that have become standard are the books of James and Kerber [40] and of Sagan [55]; [40], in particular, has a very extensive list of further references.

Chapter 9

Frobenius Groups

Results from earlier chapters join forces in this chapter to establish many of the important properties of a class of groups called Frobenius groups, which are originally defined as permutation groups but have numerous equivalent descriptions.

Frobenius groups occur naturally in a wide variety of group-theoretical settings. For example, they played a prominent role in the proof by Feit and Thompson [24] that groups of odd order are solvable.

9.1 Frobenius Groups and Their Characters

A transitive permutation group G on a set S (with $|S| > 1$) is called a *Frobenius group* if $\mathrm{Stab}_G(s) \neq 1$ for each $s \in S$, but each $g \neq 1$ in G has at most one fixed point, or equivalently $\mathrm{Stab}_G(s_1) \cap \mathrm{Stab}_G(s_2) = G_{s_1, s_2} = 1$ for $s_1 \neq s_2$ in S.

For a class of examples let F be a finite field with more than two elements and let $G = \mathrm{Aff}(F)$, the (one-dimensional) *affine group* on F, i.e.

$$G = \{\tau_{b,a} : a \in F^*, \, b \in F\},$$

where $\tau_{b,a} : F \to F$ is defined by $\tau_{b,a}(x) = ax + b$, and the group operation is composition of functions (see Chapter 2). Then G acts transitively on $S = F$, and it is easy to check that each $\tau_{b,1}$, $b \neq 0$, has no fixed points, whereas if $a \neq 1$ then $\tau_{b,a}$ has the unique fixed point $b/(1-a)$. Thus G is a Frobenius group. The translation subgroup $M = \{\tau_{b,1} : b \in F\}$ is normal in G; it is isomorphic with the additive group F. The subgroup $H = \mathrm{Stab}_G(0) = \{\tau_{0,a} : a \in F^*\}$ is isomorphic with the multiplicative group F^*, and G is the semidirect product $M \rtimes H$.

The next proposition is the first of several purely group-theoretical characterizations of Frobenius groups.

Proposition 9.1.1 *A group G is a Frobenius group if and only if it has a proper subgroup $H \neq 1$ such that $H \cap H^x = 1$ for all $x \in G \setminus H$.*

Proof $(\Rightarrow)$ Say that G acts on S, take $s \in S$, and let $H = \mathrm{Stab}_G(s)$. If $x \in G \setminus H$ then $s^x \neq s$, and if $1 \neq y \in H$ then $s^{xy} \neq s^x$ (since y fixes only s), and hence $xyx^{-1} \notin H$; i.e. $y \notin H^x$, so $H \cap H^x = 1$.

$(\Leftarrow)$ Set $S = \{Hx : x \in G\}$. Then G acts transitively on S by right multiplication. If we take $s = H \in X$ then $\mathrm{Stab}_G(s) = H$, and if $x \in G$ then $\mathrm{Stab}_G(sx) = H^x$. In particular, if $x \in G \setminus H$ then $\mathrm{Stab}_G(s) \cap \mathrm{Stab}_G(sx) = H \cap H^x = 1$, and it follows that G is a Frobenius group. $\triangle$

Corollary 9.1.2 *If G is a Frobenius group and $H \leq G$ is the stabilizer of a point then $N_G(H) = H$.*

If G is a Frobenius group on S and $H = \mathrm{Stab}_G(s)$ for some $s \in S$ then H is called a *Frobenius complement* in G. Denote by M^* the set of all $x \in G$ having *no* fixed points in S, and set $M = M^* \cup \{1\}$. Thus

$$M = (G \setminus \cup \{H^x : x \in G\}) \cup \{1\};$$

M is called the *Frobenius kernel* of G. Note that $M^x = M$ for all $x \in G$.

Proposition 9.1.3 *Suppose that G is a Frobenius group with complement H and kernel M. Then*

1. *$|M| = [G : H] > 1$, and*
2. *if $K \lhd G$ with $K \cap H = 1$ then $K \subseteq M$.*

Proof (1) Since $H = N_G(H)$ there are $[G : H]$ distinct conjugates of H, so $|\cup\{H^x : x \in G\}| = [G : H](|H| - 1) + 1 = |G| - [G : H] + 1$. Thus $|M| = |G| - (|G| - [G : H] + 1) + 1 = [G : H]$.

(2) Since $K \cap H = 1$ and $K \lhd G$ we have $K \cap H^x = 1$ for all $x \in G$ and hence $K \subseteq M$. $\triangle$

Proposition 9.1.4 *Suppose that G is a Frobenius group with complement H, and suppose $\theta \in \mathrm{cf}(H)$, with $\theta(1) = 0$. Then $(\theta^G)|_H = \theta$.*

Proof Note first that $\theta^G(1) = [G : H]\theta(1) = 0 = \theta(1)$. If $1 \neq y \in H$ then $\theta^G(y) = |H|^{-1} \sum_{t \in G} \dot\theta(t^{-1}yt)$, and $t^{-1}yt \in H$ if and only if $y \in H \cap {}^tH$ if and only if $t \in H$, in which case $\dot\theta(t^{-1}yt) = \theta(y)$, so $\theta^G(y) = \theta(y)$. $\triangle$

Recall from page 123 that the *character ring* of a group G is

$$\mathrm{Char}(G) = \left\{ \sum_{i=1}^{k} n_i \chi_i : \chi_i \in \mathrm{Irr}(G), \, n_i \in \mathbb{Z} \right\}.$$

Theorem 9.1.5 (Frobenius) *If G is a Frobenius group with complement H and kernel M then M is a normal subgroup of G.*

Proof Note that the issue here is not normality but rather the fact that M is a subgroup. We shall show that M is the intersection of the kernels of certain absolutely irreducible characters of G. Take $\varphi \neq 1_H$ in $\mathrm{Irr}(H)$, and set $\theta = \varphi - \varphi(1)1_H \in \mathrm{Char}(H)$, so $\theta(1) = 0$. Then $(\theta^G, \theta^G) = (\theta, (\theta^G)|_H) = (\theta, \theta)$, by Proposition 9.1.4, and

$$(\theta, \theta) = (\varphi - \varphi(1)1_H, \varphi - \varphi(1)1_H) = 1 + \varphi(1)^2.$$

Also $(\theta^G, 1_G) = (\theta, 1_H) = -\varphi(1)$ Thus, if we set $\varphi^* = \theta^G + \varphi(1)1_G$, then $\varphi^* \in \mathrm{Char}(G)$ and $(\varphi^*, 1_G) = 0$. Furthermore

$$(\varphi^*, \varphi^*) = 1 + \varphi(1)^2 - 2\varphi(1)^2 + \varphi(1)^2 = 1,$$

so either φ^* or $-\varphi^*$ is in $\mathrm{Irr}(G)$. Apply Proposition 9.1.4 again to see that if $y \in H$ then
$$\varphi^*(y) = \theta^G(y) + \varphi(1) = \theta(y) + \varphi(1) = \varphi(y),$$
and hence $\varphi^*|_H = \varphi$. In particular, $\varphi^*(1) = \varphi(1) > 0$, and $\varphi^* \in \mathrm{Irr}(G)$.

Set $K = \cap\{\ker(\varphi^*) : 1_H \neq \varphi \in \mathrm{Irr}(H)\} \lhd G$. If $y \in K \cap H$ then $\varphi(y) = \varphi^*(y) = \varphi^*(1) = \varphi(1)$ for all $\varphi \neq 1_H$ in $\mathrm{Irr}(H)$, and hence $y = 1$ by Corollary 5.2.31. Consequently $K \subseteq M$ by Proposition 9.1.3. On the other hand, if $1 \neq x \in M$ then $x \notin H^z$ for any $z \in G$, so

$$\varphi^*(x) = \theta^G(x) + \varphi(1) = \varphi(1) = \varphi^*(1),$$

and $x \in \ker(\varphi^*)$. Thus $K = M \lhd G$. $\qquad\qquad \triangle$

Corollary 9.1.6 *G is the semidirect product $M \rtimes H$.*

Remark Frobenius proved the theorem above in 1901. Attempts have been made to prove the theorem without using character-theoretic methods, but to date no one has succeeded. Group-theoretical proofs have been given under the extra assumption that H is solvable, by techniques involving the transfer and cohomology theory.

Exercise

If p and q are primes and G is a nonabelian group of order pq, show that G is a Frobenius group; find its kernel and a complement.

Proposition 9.1.7 *Suppose that G is a Frobenius group with complement H and kernel M. If $1 \neq x \in M$ then $C_G(x) \leq M$.*

Proof First suppose that $h \in H \cap C_G(x)$. Then $h^x = h \in H \cap H^x = 1$, so $h = 1$. In general, if $y \in C_G(x)$ but $y \notin M$, then $1 \neq y \in H^z$ for some $z \in G$, and so $^zy \in H \cap C_G(^zx) = 1$, and $y = 1$, a contradiction.   △

Proposition 9.1.8 *Suppose that G is a Frobenius group with complement H and kernel M. Then $|H| \big| |M| - 1$.*

Proof The complement H acts by conjugation on M, and if $1 \neq x \in M$ then $\text{Stab}_H(x) = C_H(x) = 1$ by Proposition 9.1.7. Thus $M \setminus \{1\}$ is a union of H-orbits, each of size $|H|$.   △

Recall A subgroup K of G is called a *Hall subgroup* if $|K|$ and $[G{:}K]$ are relatively prime.

Corollary 9.1.9 *The complement H is a Hall subgroup of G, and the kernel M is a normal Hall subgroup.*

Proposition 9.1.10 *If G is Frobenius with complement H of even order then the kernel M is abelian.*

Proof Choose $h \in H$ with $|h| = 2$. If $1 \neq x \in M$ then $x \neq x^h \in M$ by Proposition 9.1.7. Take $x, y \in M$ and suppose that $x^h \cdot x^{-1} = y^h \cdot y^{-1}$. Then $(y^h)^{-1}x^h = (y^{-1}x)^h = y^{-1}x$, so $y = x$, again by Proposition 9.1.7. Thus the map $x \mapsto x^h \cdot x^{-1}$ is 1–1 on M to M, and $M = \{x^h \cdot x^{-1}\}$. Note that $(x^h \cdot x^{-1})^h = x(x^h)^{-1} = (x^h \cdot x^{-1})^{-1}$. Thus the automorphism $z \mapsto z^h$ of M coincides with the map $z \mapsto z^{-1}$, and so M must be abelian.   △

Proposition 9.1.11 *Suppose that $G = M \rtimes H$ is a Frobenius group with kernel M and complement H, and let z be an involution in H. Then $^zx = x^{-1}$ for all $x \in M$.*

Proof Since $|H|$ is even M is abelian by Proposition 9.1.10. If $x \in M$ then $z(xz)^2 = (zxz)xz = x(zxz)z = (xz)^2z$, i.e. z centralizes $(xz)^2$. But $(xz)^2 \in M$, so $(xz)^2 = 1$, or equivalently $^zx = x^{-1}$, by Proposition 9.1.7.   △

Set $S = \{1, 2, \ldots, n\}$ and $X = S \times S$. If σ is a permutation of X and $A = [a_{ij}]$ is an $n \times n$ matrix over a field F, define $A^\sigma = [b_{ij}]$, where $b_{ij} = a_{kl}$, with $(k, l) = (i, j)^\sigma$. Check that if τ is another permutation of X then $A^{\sigma\tau} = (A^\sigma)^\tau$, so any permutation action on X determines a permutation action on the set of $n \times n$ matrices.

Proposition 9.1.12 (Brauer's Lemma) *Suppose that G is a permutation group on $X = S \times S$ as above, that $F \subseteq \mathbb{C}$, and that A is an invertible $n \times n$ matrix over F. Suppose further that for each $\sigma \in G$ the matrix A^σ can be obtained from A either by permuting the rows of A or by permuting the columns of A, so G can be viewed either as a permutation group G_r on the set of rows of A or G_c on the set of columns of A. Then the permutation characters θ_r and θ_c of G_r and G_c are equal.*

Proof If $\sigma \in G$ then there are permutation matrices $R(\sigma)$ and $C(\sigma)$ for which $R(\sigma)A = A^\sigma = AC(\sigma)$; in fact, $\sigma \mapsto R(\sigma)^t$ and $\sigma \mapsto C(\sigma)$ are the permutation representations of G_r and G_c, respectively. Thus $A^{-1}R(\sigma)A = C(\sigma)$, so tr $R(\sigma)^t =$ tr $R(\sigma) =$ tr $C(\sigma)$, all $\sigma \in G$, and $\theta_r = \theta_c$. $\triangle$

Corollary 9.1.13 *In the setting of Brauer's Lemma the numbers of orbits of G_r and G_c are equal.*

Proof Since $\theta_r = \theta_c$ we have $(\theta_r, 1_G) = (\theta_c, 1_G)$, and these are the numbers of orbits by the Orbit Formula (Proposition 3.1.2). $\triangle$

A fairly typical application of Brauer's Lemma appears in the proof of the next proposition. It should be remarked that any character table is an invertible matrix by the Second Orthogonality Relation (5.2.17).

Proposition 9.1.14 *If G is a Frobenius group with kernel M and if $\varphi \neq 1_M$ in $\mathrm{Irr}(M)$ then φ has inertia group $I_G(\varphi) = M$.*

Proof If A is the character table of M then G acts on rows of A by conjugating characters and on columns of A by conjugating M-conjugacy classes. If $x \in G$, $\chi \in \mathrm{Irr}(M)$, and L is an M-class, then the actions are related by $\chi^x(L) = \chi(^xL)$, so Brauer's Lemma (9.1.12) applies.

Choose $x \in G \setminus M$ and suppose that $^xL = L$ for some M-conjugacy class L, say $L = \mathrm{cl}_M(y)$. Thus $^xy \in L$, so $^xy = {}^my$ for some $m \in M$, $m^{-1}xy = y$, so $m^{-1}x \in C_G(y)$ and $x \in mC_G(y)$. But then $x \in M$ by Proposition 9.1.7, a contradiction unless $y = 1$. Thus $\theta_c(x) = 1$ for all $x \in G \setminus M$, and so also $\theta_r(x) = 1$ by Brauer's Lemma. But that says that $\varphi^x \neq \varphi$ if $1_M \neq \varphi \in \mathrm{Irr}(M)$. $\triangle$

If G is Frobenius with kernel M and complement H then $H \cong G/M$, so any character of H can be viewed as a character of G/M, hence also as a character of G by lifting, as discussed in Chapter 5. In particular, $\mathrm{Irr}(H)$ can be viewed as a subset of $\mathrm{Irr}(G)$ in a natural way.

Theorem 9.1.15 *Suppose that G is a Frobenius group with complement H and kernel M.*

1. If $1_M \neq \varphi \in \mathrm{Irr}(M)$ then $\varphi^G \in \mathrm{Irr}(G)$.

2. $\mathrm{Irr}(G) = \mathrm{Irr}(H) \dot\cup \{\varphi^G : 1_M \neq \varphi \in \mathrm{Irr}(M)\}$.

3. If $\chi = \varphi^G$, as in part 1, then $\chi|_H = \varphi(1)\rho_H$, where ρ_H is the regular character of H.

Proof (1) Proposition 9.1.14 and Corollary 6.4.4.

(2) Take $\chi \in \mathrm{Irr}(G)$ and let $\varphi \in \mathrm{Irr}(M)$ be a constituent of $\chi|_M$. If $\varphi \neq 1_M$ then $(\varphi^G, \chi) = (\varphi, \chi|_M) \geq 1$, so $\chi = \varphi^G$ since both are irreducible. If $\varphi = 1_M$ then $M \leq \ker(\chi)$ by Corollary 6.3.4, so we may view χ as a character of G/M, hence of H.

(3) Take $h \neq 1$ in H. If $x \in G$ then $h^x \notin M$, since $M \lhd G$. Thus

$$\chi(h) = \varphi^G(h) = |M|^{-1} \sum_{x \in G} \dot\varphi(x^{-1}hx) = 0,$$

whereas $\chi(1) = \varphi^G(1) = [G: M]\varphi(1) = \varphi(1)|H|$, so $\chi|_H = \varphi(1)\rho_H$. $\triangle$

Theorem 9.1.16 *Suppose that G is Frobenius with complement H and kernel M, and that $\varphi, \theta \in \mathrm{Irr}(M) \setminus \{1_M\}$. Then $\varphi^G = \theta^G$ if and only if $\theta \in \mathrm{Orb}_H(\varphi)$. Furthermore, $|\mathrm{Orb}_H(\varphi)| = |H|$, so G has $[c(M) - 1]/|H|$ distinct irreducible characters of the form φ^G, $\varphi \in \mathrm{Irr}(M)$.*

Proof For the first statement see the exercise on page 143. For the second use Proposition 9.1.14 and the fact that $[G: M] = |H|$. $\triangle$

Corollary 9.1.17 *If G is Frobenius with complement H and kernel M then $c(G) = c(H) + [c(M) - 1]/|H|$.*

Exercise

Use the theorems above to compute the character table of the affine group $\mathrm{Aff}(K)$ if K is a field with $q > 2$ elements.

9.2   Structure of Frobenius Groups

We begin with two more group-theoretical characterizations of Frobenius groups.

Theorem 9.2.1 *A finite group G is Frobenius if and only if it has a non-trivial proper normal subgroup M such that if $1 \neq x \in M$ then $C_G(x) \leq M$.*

Proof ($\Rightarrow$) Proposition 9.1.7.

($\Leftarrow$) We show first that M is a Hall subgroup of G. If not there is a prime p, a Sylow p-subgroup P of M, and a Sylow p-subgroup Q of G with $1 \neq P \leq Q$ but $P \neq Q$. Choose $x \in Z(Q)$, with $|x| = p$, and note that $x \notin P$, for otherwise $C_G(x) \geq Q$. Choose $y \neq 1$ in P. Then $x \in C_G(y)$, but $x \notin M$ since $P = M \cap Q$. That is a contradiction to $C_G(y) \leq M$, so M is a normal Hall subgroup of G. By the Schur-Zassenhaus Theorem (4.2.3) there is a complement H to M in G.

If $x \in G$ and $H \cap H^x \neq 1$ write $x = hz$, with $h \in H$ and $z \in M$. Then $H^x = H^{hz} = H^z$ and $H \cap H^z \neq 1$. Choose $y \neq 1$ in H with ${}^z y \in H$. Then ${}^z y y^{-1} = z(yz^{-1}y^{-1}) \in H \cap M = 1$, so $y \in C_G(z)$, contradicting $C_G(z) \leq M$ unless $z = 1$. Thus $x = h \in H$ and G is Frobenius by Theorem 9.1.1. $\triangle$

Theorem 9.2.2 *Suppose that $|G| = mn$, with $(m,n) = 1$, that either $x^m = 1$ or $x^n = 1$ for all $x \in G$, and that $M = \{x \in G \colon x^m = 1\} \triangleleft G$. Then G is Frobenius with kernel M. Conversely, if G is Frobenius with kernel M and complement H, and if $|M| = m$, $|H| = n$, then either $x^m = 1$ or $x^n = 1$ for all $x \in G$ and $M = \{x \in G \colon x^m = 1\}$.*

Proof Note first that $(|M|, n) = 1$, since $\exp(M)$ is at least a divisor of m. Since $|M| \mid |G| = mn$ we may conclude that $|M| \mid m$. But for each prime p dividing m there is a Sylow p-subgroup of G in M, so $|M| = m$ and M is a normal Hall subgroup of G. By the Schur-Zassenhaus Theorem (4.2.3) there is a complement H to M in G. If $H \cap H^x \neq 1$ but $x \notin H$ we may assume that $x \in M$. Choose $h \neq 1$ in $H \cap H^x$. Then $^x h h^{-1} = x(hx^{-1}h^{-1}) \in H \cap M = 1$, so $hx = xh$. Also $|x| \mid m$ and $|h| \mid n$, so $|xh| = |x||h|$, contradicting the fact that $|xh|$ must be a divisor of either m or n. Thus in fact H must be a Frobenius complement by Theorem 9.1.1.

The converse is clear from the definition of the Frobenius kernel. $\triangle$

Exercise

If G is Frobenius with kernel M show that M char G; show in fact that $\sigma(M) \leq M$ for every endomorphism σ of G, so M is *fully characteristic* in G.

Proposition 9.2.3 *Suppose that G is Frobenius with kernel M and complement H and that $1 \neq M_1 \leq M$, $1 \neq H_1 \leq H$, with $H_1 \leq N_G(M_1)$. Then $G_1 = M_1 H_1$ is Frobenius with kernel M_1 and complement H_1.*

Proof If $1 \neq x \in M_1$ then $C_{G_1}(x) \leq (M_1 H_1) \cap M = M_1$. Apply Theorem 9.2.1. $\triangle$

Proposition 9.2.4 *Suppose that G is Frobenius with kernel M and complement H, and that $K \leq M$, $K \neq M$, and $K \triangleleft G$. Then G/K is Frobenius, with kernel M/K.*

Proof Clearly M/K is a normal Hall subgroup of G/K. If $|M/K| = m$ and $[G \colon M] = n$, then $M/K = \{xK \in G/K \colon (xK)^m = 1\}$ by the exercise on page 84, and if $xK \notin M/K$ then $x^n = 1$, so $(xK)^n = 1$ in G/K. Apply Theorem 9.2.2. $\triangle$

Exercises

1. If G is Frobenius with kernel M and $K \triangleleft G$ show that either $K \leq M$ or $M \leq K$.

2. If G is Frobenius with kernel M and $K \leq G$ show that (1) $K \leq M$, or (2) $K \cap M = 1$, or (3) K is Frobenius with kernel $K \cap M$.

Proposition 9.2.5 *If G is abelian, and the only characteristic subgroups in G are 1 and G, then G is elementary abelian.*

Proof Choose a prime divisor p of $|G|$. The subgroup

$$H = \{x \in G : x^p = 1\}$$

is characteristic, so $H = G$ and G is elementary abelian.   △

A Frobenius group G is said to be *minimal* if no proper subgroup of G is a Frobenius group.

Theorem 9.2.6 *If G is a minimal Frobenius group with kernel M and complement H then M is elementary abelian and H has prime order.*

Proof By Proposition 9.2.3 $|H|$ is prime, say $|H| = q$. If P is a Sylow p-subgroup of M and $N = N_G(P)$ then $G = NM$ by the Frattini Argument (Proposition 4.2.1), and hence $q \mid |N|$ since $|G| = q|M|$. Since H is a Sylow q-subgroup of G we may assume that $H \leq N$. Thus HP is Frobenius by Proposition 9.2.3, so $HP = G$ by minimality of G, and we may conclude as well that $M = P$. If $K \neq 1$ is a characteristic subgroup of M then $K \triangleleft G$, and HK is Frobenius, again by 9.2.3, so $K = M$ by minimality of G. In particular, $Z(M) = M$ and M is abelian. Thus M $(= P)$ is an elementary abelian p-group by Proposition 9.2.5.   △

Theorem 9.2.7 *If G is Frobenius with complement H then no subgroup of H is Frobenius.*

Proof If the result is false we may assume, by Proposition 9.2.3 and Theorem 9.2.6, that H itself is Frobenius and minimal, hence that its kernel K is elementary abelian and its complement Q is cyclic of prime order q. If r is a prime, $r \mid |M|$, R is a Sylow r-subgroup of M, and $N = N_G(R)$, then $G = NM$ by the Frattini argument, so $|H| \mid |N|$. Since $N \cap M$ is a normal Hall subgroup of N it has a complement L by the Schur-Zassenhaus Theorem (4.2.3). Note that $G/M = NM/M \cong N/(N \cap M)$, so $|L| = [G:M] = |H|$. Thus $G = ML$, $L \cong G/M \cong H$, and L is minimal Frobenius. Since $L \leq N = N_G(R)$, RL is a subgroup of G, and if $1 \neq x \in R$, then $C_{RL}(x) \leq (RL) \cap M = R$, so RL is Frobenius with kernel R.

In other words, we could have assumed at the outset that M is an r-group, so we proceed now to do so. Furthermore we may assume by Proposition 9.2.4 that no nontrivial normal subgroups of G are properly contained in M. In particular, M has no nontrivial proper characteristic subgroups, so M is an elementary abelian r-group by Proposition 9.2.5. We may now view M (written additively) as a vector space over $\mathbb{Z}_r$. The action of H on M by conjugation is then a $\mathbb{Z}_r$-representation T of H on M. If $h \in H$ and $T(h) = 1$ then $h \in C_G(M)$, so $h = 1$ by Proposition 9.1.7, and T is faithful. The only

T-invariant subspaces are 0 and M, since M has no other subgroups normal in G; i.e. T is irreducible. Choose a finite extension F of $\mathbb{Z}_r$ that is a splitting field for both H and K (5.1.12).

Say that $T^F \sim S_1 \oplus \cdots \oplus S_l$, with each S_i absolutely irreducible, and say that $S = S_1$ acts on the F-subspace V of M^F. Write $V = V_1 \oplus \cdots \oplus V_k$, with each V_k an irreducible K-invariant subspace. But K is abelian, so each V_i is one-dimensional, and if $x \in K$ then $\widehat{S}(x)$ is a diagonal matrix. Combine the subspaces V_i so that $V = W_1 \oplus \cdots \oplus W_u$, where $\widehat{S}(x)$ restricts to a scalar matrix on each W_i, all $x \in K$, with different scalars for some x if $i \neq j$. Observe that if $v \in V$ and $\widehat{S}(x)v = \lambda_x v$, $\lambda_x \in F$, for all $x \in K$, then $v \in W_i$ for some i.

Say that $Q = \langle y \rangle$ and choose $v \in W_i$. For each $x \in K$ we have $\widehat{S}(x)\widehat{S}(y)v = \widehat{S}(y)\widehat{S}(y^{-1}xy)v = \widehat{S}(y)\lambda_{x^y}v = \lambda_{x^y}\widehat{S}(y)v$, so $\widehat{S}(y)v \in W_j$ for some j. Thus for each i there is some $j = j(i)$ so that $\widehat{S}(y)W_i = W_{j(i)}$. In other words, Q acts as a permutation group on the set $\{W_i\}$, and it has no fixed points since a fixed point would be both Q-invariant and (see above) K-invariant as a subspace, hence H-invariant, whereas S is irreducible on V. Since Q is cyclic, of prime order q, $\widehat{S}(y)$ permutes each Q-orbit in $\{W_i\}$ cyclically as a q-cycle. Relabel if necessary so that $\widehat{S}(y)W_1 = W_2$, $\widehat{S}(y)W_2 = W_3$, $\ldots$, $\widehat{S}(y)W_{q-1} = W_q$. Choose $w \neq 0$ in W_1 and set $v = w + \widehat{S}(y)w + \widehat{S}(y^2)w + \cdots$. Then clearly $\widehat{S}(y)v = v \neq 0$, so 1 is an eigenvalue of $S(y)$, hence of $T(y)$. That means that for some $z \in M$, with $z \neq 1$ (i.e. z nonzero in the vector space M) we have $T(y)z = {}^{y}z = z$. But then $y \in C_G(z)$, contradicting Proposition 9.1.7 and proving the theorem. △

Theorem 9.2.8 *The Frobenius kernel of a Frobenius group is unique; i.e. there do not exist distinct nontrivial proper normal subgroups each of which contains the centralizer of each of its nonidentity elements.*

Proof Suppose that M_1 and M_2 are both Frobenius kernels in G, say with $|M_i| = m_i$ and $[G : M_i] = n_i$. Set $N = M_1 \cap M_2$. Then $(M_1 M_2)/N = (M_1/N) \times (M_2/N)$ (internal direct product). If $M_1 \not\leq M_2$ then there is a prime p with $p \mid m_1$, $p \nmid m_2$, and if $M_2 \not\leq M_1$ there is a prime q with $q \nmid m_1$, $q \mid m_2$, so $q \mid n_1$. In that case choose $xN \in M_1/N$ of order p and $yN \in M_2/N$ of order q. Then $(xyN)^{m_1} = y^{m_1}N \neq N$ and $(xyN)^{n_1} = x^{n_1}N \neq N$, contradicting Theorem 9.2.2, according to which either $(xy)^{m_1} = 1$ or $(xy)^{n_1} = 1$. We may assume, then, that $M_1 \leq M_2$. Let H_1 be a complement for M_1, and set $K = H_1 \cap M_2 \lhd H_1$. If $1 \neq x \in K$ then $C_{H_1}(x) \leq H_1 \cap M_2 = K$, so H_1 is Frobenius with kernel K by Theorem 9.2.1. This contradicts Theorem 9.2.7, so $K = H_1 \cap M_2 = 1$. But then $|M_2| \leq [G : H_1] = |M_1|$, and hence $M_1 = M_2$. △

Remark Since M is a normal Hall subgroup it follows from the remark following the proof of the Schur-Zassenhaus Theorem (4.2.3) that the Frobenius complement H is uniquely determined up to conjugacy in G. This does not depend on the Feit-Thompson Theorem ([24]), since the kernel M must in fact be solvable (see below).

Proposition 9.2.9 (Burnside) *Suppose that G is Frobenius with kernel M and complement H, and that p, q are primes in* N. *If $K \leq H$ and $|K| = pq$ then K is cyclic.*

Proof If $p \neq q$ this follows from Theorem 9.2.7 and Exercise 3 on page 173. Suppose then that $p = q$. Thus if K of order p^2 is not cyclic then it is elementary abelian. By the reductive arguments used in the proof of Theorem 9.2.7 we may assume that $H = K$ and that M is an elementary abelian r-group for some prime $r \neq p$. As in that proof the action of H on M by conjugation determines a faithful irreducible $\mathbb{Z}_r$-representation T of H on M. Choose a finite extension F of $\mathbb{Z}_r$ that is a splitting field for H, so $T^F \sim T_1 \oplus \cdots \oplus T_k$, with $\deg T_i = 1$, all i. Thus for an appropriate choice of basis $\widehat{T}(x)$ is diagonal, all $x \in H$, and each diagonal entry $\widehat{T_i}(x)$ is a pth root of unity in F. There are at most p distinct pth roots of unity in F, and $|H| = p^2$, so there are $x, y \in H$, $x \neq y$, with $\widehat{T_1}(x) = \widehat{T_1}(y)$, or $\widehat{T_1}(xy^{-1}) = 1$. If $z = xy^{-1}$ this says that 1 is an eigenvalue of $T(z)$, so there is an eigenvector $u \in M$, $u \neq 0$ (i.e. multiplicatively $u \neq 1$), and $T(z)u = u$, or $^zu = u$. Thus $1 \neq z \in C_G(u) \cap H$, contradicting Proposition 9.1.1. $\triangle$

Theorem 9.2.10 *Suppose that G is Frobenius with complement H, that $p \,\big|\, |H|$ is prime, and that P is a Sylow p-subgroup of H. If p is odd then P is cyclic; if $p = 2$ then P is either cyclic or a generalized quaternion group Q_{2^k}, of order 8 or greater.*

Proof Take $K \leq Z(P)$ with $|K| = p$. If there is a subgroup L of P with $|L| = p$ and $L \neq K$ then KL is noncyclic of order p^2, contradicting Proposition 9.2.9. Thus P has only one subgroup of order p, and the theorem follows from Theorem 4.3.11. $\triangle$

  If a Sylow 2-subgroup of a Frobenius complement H is cyclic then it follows from the theorem above and Theorem 4.1.17 that H is split metacyclic, i.e. a split extension of a cyclic group by another cyclic group. In particular, H is solvable. Zassenhaus has shown more generally (see [50], page 196) that if H is a solvable Frobenius complement then H has a normal subgroup H_0 which is split metacyclic and such that the quotient H/H_0 is isomorphic with a subgroup of the symmetric group S_4.
  There are nonsolvable Frobenius complements, however. An example with $H = \mathrm{SL}(2,5)$ will follow shortly. In this case Zassenhaus has shown (see [50], page 204) that it is approximately the only such example, in that if H is a nonsolvable Frobenius complement then H has a subgroup H_0 of index 1 or 2 with $H_0 = \mathrm{SL}(2,5) \times K$, where K is split metacyclic of order prime to 30.

For the example we observe first that $H = \mathrm{SL}(2,5)$ has the presentation

$$\langle x, y, z \mid x^3 = y^5 = z^2 = 1,\ x^z = x,\ y^z = y,\ (xy)^2 = z\rangle,$$

via

$$x \mapsto \begin{bmatrix} 0 & 1 \\ 4 & 4 \end{bmatrix},\quad y \mapsto \begin{bmatrix} 1 & 0 \\ 1 & 1 \end{bmatrix},\quad z \mapsto \begin{bmatrix} -1 & 0 \\ 0 & -1 \end{bmatrix}.$$

The matrices satisfy the relations in the presentation and they generate $SL(2,5)$, so there is a homomorphism onto $\mathrm{SL}(2,5)$. A Todd-Coxeter coset enumeration (see Chapter 1) of the subgroup $\langle y, z\rangle$ shows that it has index 12, so the group presented has order 120 or less and hence is isomorphic with H.

Let $F = \mathbb{Z}_{29}$ and note that $12^2 = -1$, $11^2 = 5$ (any finite field of characteristic not 2, 3, or 5 in which -1 and 5 are squares would do as well). Define

$$X = \begin{bmatrix} -1 & 1 \\ -1 & 0 \end{bmatrix},\quad Y = \begin{bmatrix} 0 & -12 \\ -12 & -6 \end{bmatrix},\quad Z = \begin{bmatrix} -1 & 0 \\ 0 & -1 \end{bmatrix}$$

in $\mathrm{SL}(2,29)$. Verify that the matrices X, Y, Z satisfy the relations in the presentation for H. It follows that H maps isomorphically into $\mathrm{SL}(2,29)$ via $x \mapsto X$, $y \mapsto Y$, $z \mapsto Z$. With only a slight abuse of notation we take the point of view that H *is* that subgroup of $\mathrm{SL}(2,29)$.

Now set $M = F \oplus F$, so $H \leq \mathrm{Aut}(M)$, and finally let G be the resulting semidirect product $M \rtimes H$.

Exercise

If $v \in M$, but $v \neq \begin{bmatrix} 0 \\ 0 \end{bmatrix} = 1_M$, show that $C_G(v) = M$. Thus G is Frobenius, with complement $H = \mathrm{SL}(2,5)$. (*Hint* If $(w,g) \in C_G(v) \setminus M$ then v is an eigenvector of g for eigenvalue 1. But then both eigenvalues of g are 1 (since $\det g = 1$), and the order of g is 29.)

If G is any group and if $\sigma \in \mathrm{Aut}(G)$ fixes only 1, then σ is called a *fixed-point-free* (abbreviated *f.p.f.*) automorphism. It was known to Frobenius that any (finite) group with an f.p.f. automorphism of order 2 or 3 must be nilpotent, and he conjectured that any group with an f.p.f. automorphism of prime order must be nilpotent.

If G is Frobenius then every $x \neq 1$ in H provides an f.p.f. automorphism of M by conjugation, since x would be in the centralizer of any fixed point. In particular, M has f.p.f. automorphisms of prime order, which is the basis for the conjecture by Frobenius that M must be nilpotent. J. Thompson, in his Ph.D. dissertation in 1959, proved the more general Frobenius conjecture: a group with an f.p.f. automorphism of prime order is nilpotent. For a proof see [29], page 337.

It seems in fact that in Frobenius's time no examples were known of nonabelian Frobenius kernels. L. Weisner thought in 1939 that he had proved all Frobenius kernels are abelian, but in 1940 O. J. Schmidt produced an example of a nonabelian kernel. We will give a later example due to N. Ito.

Take $p, n \in \mathbb{N}$, with p prime, n odd, and $n > 1$, and let F be the field with p^n elements. For each $a, b \in F$ define a matrix

$$\mu(a,b) = \begin{bmatrix} 1 & a & b \\ 0 & 1 & a^p \\ 0 & 0 & 1 \end{bmatrix}.$$

Since the map $a \mapsto a^p$ is an automorphism of F it is easy to check that

$$\mu(a,b)\mu(c,d) = \mu(a + c, ac^p + b + d),$$

and it follows that $M = \{\mu(a,b) : a, b, \in F\}$ is a nonabelian group of order p^{2n}.

Choose $r \in F$ of (multiplicative) order $(p^n - 1)/(p - 1)$, and define a map $\sigma : M \to M$ via $\,^\sigma\mu(a,b) = \mu(ra, r^{p+1}b)$. Again it is easy to check that σ is an automorphism of M, so $H = \langle \sigma \rangle \leq \mathrm{Aut}(M)$. Define G to be the corresponding semidirect product $M \rtimes H$.

Exercises

1. Show that if $1 \neq x = \mu(a,b) \in M$ (above) then $C_G(x) \leq M$, and conclude that G is a Frobenius group with nonabelian kernel M.

2. Let $F = \mathbb{Z}_5$ and $H = \langle a, b \rangle \leq \mathrm{SL}(2,5)$, where

$$a = \begin{bmatrix} 3 & 4 \\ 3 & 1 \end{bmatrix} \quad \text{and} \quad b = \begin{bmatrix} 3 & 0 \\ 0 & 2 \end{bmatrix}.$$

Set $M = F \oplus F$, so $H \leq \mathrm{Aut}\, M$, and set $G = M \rtimes H$.

   (a) Show that $H \cong Q_2 \rtimes C_3$, of order 24. (*Hint* Consider $\langle b, b^a \rangle$.)

   (b) Show that if $1 \neq x \in M$ (i.e. $x \neq \begin{bmatrix} 0 \\ 0 \end{bmatrix}$) then $C_G(x) \leq M$. (*Hint* It is sufficient to show that if $1 \neq c \in H$ then c does not have 1 as an eigenvalue.)

   (c) Conclude that G is Frobenius with complement H and that H is solvable and nonabelian.

Chapter 10

Splitting Fields

This chapter explores various aspects of the effects on characters and representations of a finite group when the ground field is extended. The particular case of $\mathbb{R}$ extended to $\mathbb{C}$ receives special attention, with interesting consequences relating to elements of order 2.

10.1 Splitting

Recall that the *exponent* of a finite group G is the least common multiple of the orders of all the elements of G; it is the minimal integer $m \geq 1$ for which $x^m = 1$ for all $x \in G$. Denote the exponent of G by $\exp(G)$.

It seems to have been conjectured by H. Maschke in about 1900 that if $\exp(G) = m$ and $\zeta \in \mathbb{C}$ is a primitive mth root of unity then $F = \mathbb{Q}(\zeta)$ is a splitting field for G. The conjecture was proved by R. Brauer [6] in 1945.

The next theorem should probably be called Blichfeldt's Splitting Field Theorem. It will be subsumed immediately by the more general theorem of Brauer; it is included because its proof requires much less preparation but also for historical interest – it provided an early proof of a special case of Maschke's conjecture..

Theorem 10.1.1 *Suppose that G is nilpotent of exponent m. Let $\zeta = \zeta_m \in \mathbb{C}$ be a primitive mth root of unity, and set $F = \mathbb{Q}(\zeta)$. Then F is a splitting field for G.*

Proof Recall from Corollary 6.3.11 that if $\chi \in \mathrm{Irr}(G)$ then $\chi = \lambda^G$ for some *linear* character λ of a subgroup $H \leq G$. It follows that if $\widehat{S}$ is a (one-dimensional) matrix representation of H with character λ then $\widehat{T} = \widehat{S}^G$ is a matrix representation of G with character χ. But $\widehat{S}(x) = \lambda(x) \in F$ for all $x \in G$, and consequently all matrix entries of $\widehat{T}(x)$ are in F by Proposition 6.2.3, i.e. $\widehat{T}$ is an F-representation. Thus a full set of inequivalent absolutely irreducible matrix representations of G can be written over F and F is a splitting field.   △

Theorem 10.1.2 (Brauer's Splitting Field Theorem) *Suppose that G is of exponent m, let $\zeta = \zeta_m \in \mathbb{C}$ be a primitive mth root of unity, and set $F = \mathbb{Q}(\zeta)$. Then F is a splitting field for G.*

Proof Let T be an irreducible matrix representation of G over $\mathbb{C}$, with character $\chi \in \mathrm{Irr}(G)$. It will suffice to show that T is equivalent with an F-representation. Suppose by way of contradiction that it is not. By Brauer's Induction Theorem (6.5.4) we may write $\chi = \theta_1 - \theta_2$, where each θ_i is a character that is a $\mathbb{Z}$-linear combination of characters induced from linear characters of subgroups of G. As in the proof of Theorem 10.1.1 there are F-matrix representations S_1 and S_2 of G having characters θ_1 and θ_2, respectively. Since $\chi + \theta_2 = \theta_1$, $T \oplus S_2$ is equivalent, over $\mathbb{C}$, to S_1 by Theorem 5.2.12. Thus, even though T is not equivalent with an F-representation, $T \oplus S_2$ is for some F-representation S_2. We may assume that S_2 is of lowest possible degree with that property. By Maschke's Theorem (5.1.1) we may write $\theta_1 = \eta_1 + \cdots + \eta_k$ and $\theta_2 = \xi_1 + \cdots + \xi_\ell$, where each η_i is the character of an irreducible F-matrix representation U_i, and likewise each ξ_i is the character of an irreducible F-matrix representation V_i. Thus

$$\chi + \xi_1 + \cdots + \xi_\ell = \eta_1 + \cdots + \eta_k.$$

Since $(\chi + \theta_2, \xi_\ell) > 0$ we have $(\theta_1, \xi_\ell) > 0$, and hence $(\eta_i, \xi_\ell) > 0$ for some i; we may assume that $(\eta_k, \xi_\ell) > 0$. By Theorem 5.2.6 (i) it follows that $\eta_k = \xi_\ell$, and so

$$\chi + \xi_1 + \cdots + \xi_{\ell-1} = \eta_1 + \cdots + \eta_{k-1}.$$

But then $T \oplus V_1 \oplus \cdots \oplus V_{\ell-1}$ is equivalent with $U_1 \oplus \cdots \oplus U_{k-1}$, contradicting the minimality of the degree of S_2 and completing the proof.   △

10.2   The Schur Index

The dihedral group $D_4 = \langle a, b \mid a^4 = b^2 = (ab)^2 = 1 \rangle$ has a $\mathbb{Q}$-matrix representation T with

$$T(a) = \begin{bmatrix} 0 & -1 \\ 1 & 0 \end{bmatrix} \quad \text{and} \quad T(b) = \begin{bmatrix} 1 & 0 \\ 0 & -1 \end{bmatrix},$$

with character χ_T whose values at 1, a^2, a, b, ab are $2, -2, 0, 0, 0$.

The quaternion group $Q_2 = \langle c, d \mid c^4 = 1, c^2 = d^2, cd = dc^{-1} \rangle$ has a character χ that "equals" χ_T, in the sense that it takes on the same values on the five classes of Q_2 (suitably ordered; see Exercise 2 on page 117). If

S is a matrix representation with character χ then $S(c)$ must have order 4 and must be diagonalizable. If $S(c)$ is to have real entries it must have eigenvalues $\pm i$ and be similar (over $\mathbb{C}$) to $\begin{bmatrix} i & 0 \\ 0 & -i \end{bmatrix}$, so all possible $S(c)$ are similar (over $\mathbb{R}$; e.g. see [31], page 143), and without loss of generality we may take $S(c) = \begin{bmatrix} 0 & -1 \\ 1 & 0 \end{bmatrix}$.

Let us try next to find $S(d) = \begin{bmatrix} x & y \\ u & v \end{bmatrix}$ with rational, or at least real, entries. Since $cd = dc^{-1}$ we have $\begin{bmatrix} -u & -v \\ x & y \end{bmatrix} = \begin{bmatrix} -y & x \\ -v & u \end{bmatrix}$, and therefore $u = y$ and $v = -x$. Since d^2 is of order 2 in the center we must have $S(d^2) = \begin{bmatrix} -1 & 0 \\ 0 & -1 \end{bmatrix}$, so

$$\begin{bmatrix} x & y \\ y & -x \end{bmatrix} \begin{bmatrix} x & y \\ y & -x \end{bmatrix} = \begin{bmatrix} x^2 + y^2 & 0 \\ 0 & x^2 + y^2 \end{bmatrix} = \begin{bmatrix} -1 & 0 \\ 0 & -1 \end{bmatrix},$$

which is of course impossible over $\mathbb{R}$.

Thus, even though the values of χ lie in $\mathbb{Q}$, χ is *not* a $\mathbb{Q}$-character, i.e. it is not the character of a $\mathbb{Q}$-representation of Q_2. If we extend $\mathbb{Q}$ to $K = \mathbb{Q}(i)$ ($i^2 = -1$ as usual), then it is easy to check that with $S(c)$ as above and $S(d) = \begin{bmatrix} i & 0 \\ 0 & -i \end{bmatrix}$ we obtain a K-representation with character χ.

We shall return to this example after a brief digression.

If F is a field and K is a finite extension of F it is customary (and useful) to view K as an F-vector space. Each $a \in K$ determines an F-linear transformation of K (call it $\tilde{a}$) by multiplication; i.e. $\tilde{a}(b) = ab$, all $b \in K$. If a basis $\{b_1, \ldots, b_n\}$ for K over F is chosen then each $\tilde{a}$ is represented by an $n \times n$ F-matrix $\check{a}$. The mapping $a \mapsto \check{a}$ is called the *regular (matrix) representation* of K over F – it is an isomorphism from K to a field of matrices over F.

For each $a \in K$ define the *trace* from K to F of a to be $\mathrm{Tr}_{K/F}(a) = \mathrm{tr}(\tilde{a}) = \mathrm{tr}(\check{a})$, which is of course independent of the basis chosen for K over F. The function $\mathrm{Tr}_{K/F} \colon K \to F$ is an F-linear transformation, and if $b \in F$ then $\mathrm{Tr}_{K/F}(b) = [K \colon F] \cdot b$ (since $\check{b} = bI$).

Returning to the example above, take $\{1, i\}$ as a basis for K over $\mathbb{Q}$ and observe that

$$\check{1} = \begin{bmatrix} 1 & 0 \\ 0 & 1 \end{bmatrix}, \quad \check{i} = \begin{bmatrix} 0 & -1 \\ 1 & 0 \end{bmatrix},$$

so for each $a = u + vi \in K$ we have

$$\check{a} = \begin{bmatrix} u & -v \\ v & u \end{bmatrix}.$$

Replace each matrix entry a in the representation S by its regular representation $\breve{a}$ to determine a $\mathbb{Q}$-representation S', with

$$
S'(c) = \left[\begin{array}{cc|cc} 0 & 0 & -1 & 0 \\ 0 & 0 & 0 & -1 \\ \hline 1 & 0 & 0 & 0 \\ 0 & 1 & 0 & 0 \end{array}\right], \quad
S'(d) = \left[\begin{array}{cc|cc} 0 & -1 & 0 & 0 \\ 1 & 0 & 0 & 0 \\ \hline 0 & 0 & 0 & 1 \\ 0 & 0 & -1 & 0 \end{array}\right].
$$

Note that the character of S' is $\chi' = 2\chi$. A routine calculation with block matrices shows that the centralizer $\mathcal{C}_{\mathbb{Q}}(S')$ consists of all rational matrices of the form

$$
\left[\begin{array}{cc|cc} x & y & v & u \\ -y & x & u & -v \\ \hline -v & -u & x & y \\ -u & v & -y & x \end{array}\right],
$$

which is Hamilton's division algebra $\mathbb{H}_{\mathbb{Q}}$ of rational quaternions (e.g. see [31], page 48), so S' is an irreducible $\mathbb{Q}$-representation by Proposition 5.1.3.

To summarize, χ is an absolutely irreducible character of $G = Q_2$ having values in $\mathbb{Q}$; it is not a $\mathbb{Q}$-character, but 2χ *is* the character of an irreducible $\mathbb{Q}$-representation.

Proposition 10.2.1 *Suppose that F and K are fields, with $F \subseteq K \subseteq \mathbb{C}$ and $[K:F]$ finite, and that T is a K-(matrix) representation of G with character χ. Then there is an F-representation T' of G with character $\chi' = \mathrm{Tr}_{K/F} \circ \chi$.*

Proof Choose a basis for K over F. If $T(x) = [t_{ij}(x)]$, $x \in G$, let $T'(x)$ be the F-matrix whose partitioned form is $[\breve{t}_{ij}(x)]$, with entries of $[t_{ij}(x)]$ replaced by their regular representation matrices for K over F. Then T' is an F-representation and

$$
\begin{aligned}
\chi'(x) &= \mathrm{tr}\, T'(x) = \sum_i \mathrm{tr}[\breve{t}_{ii}(x)] = \sum_i \mathrm{Tr}_{K/F}[t_{ii}(x)] \\
&= \mathrm{Tr}_{K/F}\Big(\sum_i t_{ii}(x)\Big) = \mathrm{Tr}_{K/F}(\chi(x)).
\end{aligned}
$$

$\triangle$

For any field $F \subseteq \mathbb{C}$ and any $\mathbb{C}$-character ψ of G define $F(\psi)$ to be the field obtained by adjoining to F all the values of ψ, i.e. $F(\psi) = F(\{\psi(x) : x \in G\})$, a finite extension of F.

Proposition 10.2.2 *If $F \subseteq \mathbb{C}$ and $\chi \in \mathrm{Irr}(G)$ then there is some $m > 0$ in $\mathbb{N}$ such that $m\chi$ is an $F(\chi)$-character.*

Proof Choose a finite extension K of $F(\chi)$ for which χ is a K-character (e.g. K could be a splitting field for G). Apply Proposition 10.2.1 to χ, so that $\mathrm{Tr}_{K/F(\chi)} \circ \chi$ is the character of an $F(\chi)$-representation of G. Set

$m = [K : F(\chi)]$. Since $\chi(x) \in F(\chi)$ we have $\mathrm{Tr}_{K/F(\chi)}(\chi(x)) = m\chi(x)$ for all $x \in G$. △

The proposition above calls for a definition. If $\chi \in \mathrm{Irr}(G)$ and $F \subseteq \mathbb{C}$ then the *Schur index* $m_F(\chi)$ of χ relative to F is the least positive integer m for which $m\chi$ is an $F(\chi)$-character, i.e. the character of an $F(\chi)$-representation of G.

Clearly $m_F(\chi)\chi$ is the character of an irreducible $F(\chi)$-representation, by the minimality of $m_F(\chi)$.

Proposition 10.2.3 *Suppose that $\chi \in \mathrm{Irr}(G)$, $F \subseteq \mathbb{C}$, and $0 < n \in \mathbb{N}$. If $n\chi$ is an $F(\chi)$-character then $m_F(\chi) \,\big|\, n$.*

Proof Say that T is an $F(\chi)$-representation of G with character $n\chi$. By Maschke's Theorem (5.1.1), $T \sim T_1 \oplus \cdots \oplus T_k$, with each T_i an irreducible $F(\chi)$-representation. If ψ_i is the character of T_i then $\psi_i = n_i\chi$, with $0 < n_i \in \mathbb{N}$ for each i and $\sum_i n_i = n$. Now $n_i\chi$ and $m_F(\chi)\chi$ are both irreducible $F(\chi)$-characters, and $(n_i\chi, m_F(\chi)\chi) = n_i m_F(\chi) > 0$, so $n_i\chi = m_F(\chi)\chi$ by Theorem 5.2.6. Thus $n = km_F(\chi)$. △

Proposition 10.2.4 *Suppose that $\chi \in \mathrm{Irr}(G)$, $F \subseteq K \subseteq \mathbb{C}$ with $[K : F]$ finite, and χ is a K-character of G. Then $m_F(\chi) \,\big|\, [K : F(\chi)]$.*

Proof Set $n = [K : F(\chi)]$ and apply Proposition 10.2.1. There is an $F(\chi)$-representation T whose character is $\mathrm{Tr}_{K/F(\chi)} \circ \chi$. Since $\chi(x) \in F(\chi)$, all $x \in G$, we have $\mathrm{Tr}_{K/F(\chi)}(\chi(x)) = [K : F(\chi)]\chi(x) = n\chi(x)$, and $n\chi$ is an $F(\chi)$-character. Thus $m_F(\chi) \,\big|\, n$ by Proposition 10.2.3. △

Remark Given $F \subseteq \mathbb{C}$ and $\chi \in \mathrm{Irr}(G)$ there is a field L, with $F(\chi) \subseteq L \subseteq \mathbb{C}$, such that χ is an L-character and $[L : F(\chi)] = m_F(\chi)$. Thus $m_F(\chi)$ also could have been defined as the minimal degree of an extension field K of $F(\chi)$ for which χ is a K-character. For a proof (which involves the theory of central simple algebras) see [18], page 469, or [22], page 128.

Suppose that F is any subfield of $\mathbb{C}$. If $\exp(G) = n$ and we set $K = F(\zeta)$, where $\zeta \in \mathbb{C}$ is a primitive nth root of unity, then K is a splitting field for G by Brauer's Splitting Field Theorem (10.1.2). The Galois group $\mathcal{G}(K : F)$ is abelian (e.g. see [31], page 97).

If θ is any $\mathbb{C}$-character of G then $F(\theta) \subseteq K$, and $F(\theta)$ is a normal extension of F since $\mathcal{G}(K : F)$ is abelian. Let $\mathcal{G} = \mathcal{G}(F(\theta) : F)$, viewed as a subgroup (via restriction) of $\mathcal{G}(K : F)$. If T is a K-matrix representation with character θ, say with $T(x) = [t_{ij}(x)]$, all x, then for each $\sigma \in \mathcal{G}$ we may define T^σ via $T^\sigma(x) = [\sigma(t_{ij}(x))]$. Then T^σ is also a K-representation of G, whose character is clearly θ^σ, with $\theta^\sigma(x) = \sigma(\theta(x))$, all $x \in G$. Note that if $1 \neq \sigma \in \mathcal{G}$ then $\theta^\sigma \neq \theta$, by elementary Galois theory. Furthermore $(\theta^\sigma, \theta^\sigma) = (\theta, \theta)$, so θ^σ is absolutely irreducible if and only if θ is.

The characters θ^σ, $\sigma \in \mathcal{G} = \mathcal{G}(F(\theta):F)$, are called the *Galois conjugates* of θ over F. Note that $\sum\{\theta^\sigma : \sigma \in \mathcal{G}\}$ has its values in F.

The fact that $\mathrm{Irr}(G)$ contains the Galois conjugates of each of its characters can be very useful in calculating character tables. For example, the final character χ_3 of A_5 obtained on page 134 via the Second Orthogonality Relation could have been written immediately as a Galois conjugate of χ_2.

The next proposition is a bit of a digression; it illustrates an easy application of the Galois conjugates discussed above.

Proposition 10.2.5 *Suppose that* $\exp(G) = n$, $x \in G$, *and that* θ *is a* $\mathbb{C}$-*character of* G. *Suppose that if* $m \in \mathbb{N}$ *is relatively prime to* n *then* $x \sim x^m$ *(conjugate in* G*). Then* $\theta(x) \in \mathbb{Q}$.

Proof Let $\mathcal{G} = \mathcal{G}(\mathbb{Q}(\zeta):\mathbb{Q})$, as above, consisting of the automorphisms σ_m determined by $\sigma_m : \zeta \mapsto \zeta^m$, with $(m,n) = 1$ and $0 \le m < n$. As in the proof of Proposition 5.2.5 we have $\theta(x) = \sum_i \zeta^{e_i}$. Since $x \sim x^m$ we have $\theta(x) = \theta(x^m) = \sum_i \zeta^{e_i m} = \sum_i \sigma_m(\zeta^{e_i}) = \sigma_m(\theta(x))$. Thus $\theta(x)$ is fixed by all $\sigma_m \in \mathcal{G}$ and hence must be in $\mathbb{Q}$. $\triangle$

Remark In fact, of course, $\theta(x) \in \mathbb{Z}$ in the proposition above, since $\theta(x)$ is also an algebraic integer.

Corollary 10.2.6 *Characters of the symmetric group* S_n *are integer-valued.*

Proof The exponent of S_n is clearly $LCM\{k \in \mathbb{N}: 1 \le k \le n\}$, so $m \in \mathbb{N}$ is relatively prime to $\exp(S_n)$ if and only if it is relatively prime to $n!$. In that case x and x^m have the same cycle structure, so $x \sim x^m$. $\triangle$

It is in fact true that all of the absolutely irreducible representations of S_n can be realized over the field $\mathbb{Q}$ of rational numbers (e.g. see [40] or [55]) i.e. all have Schur index 1 relative to $\mathbb{Q}$.

Exercise

Suppose that θ is a $\mathbb{C}$-character of G, with $\theta(x) \in \mathbb{Q}$ for all $x \in G$. If $y, z \in G$ and $\langle y \rangle = \langle z \rangle$ show that $\theta(y) = \theta(z)$. (If $|y| = n$ think about the Galois group $\mathcal{G} = \mathcal{G}(\mathbb{Q}(\zeta_n):\mathbb{Q})$.)

Back to the Schur index.

Proposition 10.2.7 *Suppose that* F *is a subfield of* $\mathbb{C}$ *and that* ψ *is an irreducible* F-*character of* G. *Then there is some* $\chi \in \mathrm{Irr}(G)$ *such that* $\psi = m_F(\chi) \sum\{\chi^\sigma : \sigma \in \mathcal{G}\}$, *where* $\mathcal{G} = \mathcal{G}(F(\chi):F)$. *Furthermore* ψ *is the unique irreducible* F-*character for which* $(\psi, \chi) \ne 0$.

Proof Choose $\chi \in \mathrm{Irr}(G)$ with $(\psi, \chi) = m \neq 0$. Then $(\psi, \chi^\sigma) = (\psi, \chi) = m$ for all $\sigma \in \mathcal{G}$, since ψ has its values in F. Thus $\psi = m \sum \{\chi^\sigma : \sigma \in \mathcal{G}\} + \psi_0$, where ψ_0 is either 0 or is a character for which $(\psi_0, \chi^\sigma) = 0$ for all $\sigma \in \mathcal{G}$. By the remark following the proof of Proposition 10.2.4 there is a field $L \supseteq F(\chi)$ such that χ is an L-character and $m_F(\chi) = [L : F(\chi)]$. Set $\varphi = m_F(\chi) \sum \{\chi^\sigma : \sigma \in \mathcal{G}\}$, and observe that then $\varphi = m_F(\chi) \mathrm{Tr}_{F(\chi)/F} \circ \chi$ (e.g. see [31], page 117), so

$$\varphi = [L : F(\chi)] \, \mathrm{Tr}_{F(\chi)/F} \circ \chi = \mathrm{Tr}_{L/F(\chi)} \circ \mathrm{Tr}_{F(\chi)/F} \circ \chi = \mathrm{Tr}_{L/F} \circ \chi,$$

and φ is an F-character by Proposition 10.2.1. Since ψ is an irreducible F-character and $(\psi, \varphi) \neq 0$ we have $\varphi = \psi + \psi_1$, with ψ_1 either 0 or a character. Thus

$$\varphi = m_F(\chi) \sum \{\chi^\sigma : \sigma \in \mathcal{G}\} = m \sum \{\chi^\sigma : \sigma \in \mathcal{G}\} + \psi_0 + \psi_1.$$

Comparing summands, we see that $m \leq m_F(\chi)$ and $\psi_0 = 0$.

Observe next that ψ and $m_F(\chi)\chi$ are both $F(\chi)$-characters and that $m_F(\chi)\chi$ is irreducible as an $F(\chi)$-character. Thus

$$m \sum \{\chi^\sigma : \sigma \in \mathcal{G}\} = \psi = m_F(\chi)\chi + \psi_2,$$

where ψ_2 is either 0 or is an $F(\chi)$-character. Comparing summands again, we see that $m_F(\chi) \leq m$, and so

$$\psi = m_F(\chi) \sum \{\chi^\sigma : \sigma \in \mathcal{G}\}$$

as desired.

For the uniqueness – if η is an irreducible F-character with $(\eta, \chi) \neq 0$ then $(\eta, \psi) \neq 0$ (since χ is a constituent of both), so $\eta = \psi$ by Theorem 5.2.6. $\triangle$

Corollary 10.2.8 *If $F \subseteq \mathbb{C}$, $\chi \in \mathrm{Irr}(G)$, and $0 < m \in \mathbb{N}$, then m is the Schur index of χ if and only if m is the multiplicity of χ as a constituent of some irreducible F-character.*

Proof It is sufficient to observe that there is some irreducible F-character ψ having χ as a constituent. $\triangle$

Theorem 10.2.9 (Schur) *Suppose that $F \subseteq \mathbb{C}$, $\chi \in \mathrm{Irr}(G)$, and η is an F-character of G having χ as a constituent. Then $m_F(\chi) \,\big|\, (\eta, \chi)$.*

Proof Let ψ be an F-irreducible summand of η having χ as a constituent, so $(\psi, \chi) = m_F(\chi)$ as in the proof of Proposition 10.2.7. Also ψ is unique by Proposition 10.2.7, so if $\eta = k\psi + \theta$, where θ is an F-character with $(\psi, \theta) = 0$, then $(\chi, \theta) = 0$ as well; hence $(\eta, \chi) = k m_F(\chi)$. $\triangle$

Corollary 10.2.10 *If $F \subseteq \mathbb{C}$ and $\chi \in \mathrm{Irr}(G)$ then $m_F(\chi) \mid \chi(1)$.*

Proof Take η to be the regular character ρ, which is an F-character, and recall from Proposition 5.2.13 that $\chi(1) = (\rho, \chi)$. $\triangle$

It follows from the corollary above and Ito's Theorem (6.3.9) that if G is a dihedral group D_n, or a quaternion group Q_n, then every $\chi \in \mathrm{Irr}(G)$ has Schur index 1 or 2.

Exercise

If $F \subseteq \mathbb{C}$ show that $m_F(\chi) = 1$ for all $\chi \in \mathrm{Irr}(D_n)$ (see page 147). (*Hint* If χ is nonlinear what is its multiplicity in 1_B^G?)

10.3 $\mathbb{R}$ versus $\mathbb{C}$

An element $x \in G$ is called a *real element* if it is conjugate to its own inverse, $x \sim x^{-1}$. A conjugacy class K of G is called a *real class* if all elements of K are real, i.e. if $K = K^{-1}$.

Proposition 10.3.1 *If $x \in G$, then x is real if and only if $\chi_i(x) \in \mathbb{R}$ for all $\chi_i \in \mathrm{Irr}(G)$.*

Proof ($\Rightarrow$) If x is real and $\chi_i \in \mathrm{Irr}(G)$ then $\chi_i(x) = \chi_i(x^{-1}) = \overline{\chi_i(x)}$ by Propositions 5.2.1 and 5.2.5.

($\Leftarrow$) If all $\chi_i(x)$ are real then $\chi_i(x) = \overline{\chi_i(x)} = \chi_i(x^{-1})$, all i, again by Proposition 5.2.5. Thus $x \sim x^{-1}$, since the set $\mathrm{Irr}(G)$ is a basis for $\mathrm{cf}(G)$ (Corollary 5.2.19). $\triangle$

Proposition 10.3.2 *The number of real-valued absolutely irreducible characters of G is equal to the number of real conjugacy classes in G.*

Proof Let A denote the character table of G, and denote by $H = \langle \sigma \rangle$ a group of order 2. Then H permutes the entries of A either by acting on rows via $\chi_i^\sigma = \overline{\chi_i}$, or by acting on columns via $^\sigma K_j = K_j^{-1}$. The two actions agree, since $\overline{\chi_i(K_j)} = \chi_i(K_j^{-1})$ by Proposition 5.2.5. By Brauer's Lemma (9.1.12) the two permutation characters θ_r and θ_c of H on the rows and columns, respectively, are equal. But $\theta_r(\sigma)$ counts fixed rows, i.e. real-valued characters, and $\theta_c(\sigma)$ counts fixed columns, i.e. real classes. $\triangle$

Theorem 10.3.3 (Burnside) *If G has odd order then 1_G is the only real-valued absolutely irreducible character of G.*

Proof By Proposition 10.3.2 it will suffice to show that $K = \{1\}$ is the only real conjugacy class in G. Suppose that $y^{-1}xy = x^{-1}$. Then $y^{-2}xy^2 = x$, so $y^2 \in C_G(x)$. Since $|G|$ is odd $\langle y \rangle = \langle y^2 \rangle$, and $y \in C_G(x)$. But then $x = x^{-1}$, so $x^2 = 1$, and $x = 1$. $\triangle$

Corollary 10.3.4 (Burnside) *If* $|G|$ *is odd then* $|G| \equiv c(G)$ *(mod* 16*).*

Proof By the theorem we may label the characters in $\mathrm{Irr}(G)$ as $\chi_0 = 1_G$, then $\chi_1, \ldots, \chi_s$; $\chi_{s+1}, \ldots, \chi_{2s}$, where $\chi_{s+i} = \overline{\chi_i}$, $1 \le i \le s$. Say that $\chi_i(1) = 2m_i + 1$, all i. Then, by Theorem 5.2.14,

$$
\begin{aligned}
|G| &= 1 + 2\sum_{i=1}^{s}(2m_i + 1)^2 = 1 + 2\sum_i (4m_i^2 + 4m_i + 1) \\
&= 1 + 2s + 8\sum_i m_i(m_i + 1) = c(G) + 8m,
\end{aligned}
$$

where $m = \sum_i m_i(m_i + 1)$ is even.   △

Exercise

If $|G|$ is odd, and if $\chi(1) \equiv 1 \pmod{2^k}$ for all $\chi \in \mathrm{Irr}(G)$, show that $|G| \equiv c(G) \pmod{2^{k+2}}$.

Recall If χ is the character of an F-representation S on a vector space U then χ^2 is the character of $T = S \otimes S$ on $V = U \otimes_F U$.

The map $f : U \times U \to V$ defined by $f(u, v) = v \otimes u$ is easily checked to be "balanced," so there is a uniquely determined and well-defined linear transformation $*$ on V with $(u \otimes v)^* = v \otimes u$, all $u, v \in U$ (e.g. see [31], page 158). The square of $*$ is the identity, so its minimal polynomial divides $x^2 - 1$. Let V_s and V_a denote the respective eigenspaces for 1 and -1, respectively, of $*$, the so-called subspaces of *symmetric* and *antisymmetric* tensors.

Thus $V_s = \{v \in V : v^* = v\}$ and $V_a = \{v \in V : v^* = -v\}$, and clearly $V_s \cap V_a = 0$. If $v \in V$ then $v + v^* \in V_s$, $v - v^* \in V_a$, and $v = \frac{1}{2}(v + v^*) + \frac{1}{2}(v - v^*)$, so $V = V_s \oplus V_a$.

Since $T(x)(v^*) = (T(x)v)^*$ for all $x \in G$, it follows that T preserves the eigenspaces of $*$, i.e. that V_s and V_a are T-invariant subspaces of V. Thus, if T_s and T_a are the restrictions of T to V_s and V_a, respectively, then $T = T_s \oplus T_a$, and we may write $\chi^2 = \chi_s + \chi_a$, the corresponding sum of characters; χ_s and χ_a are sometimes called the *symmetric square* and the *antisymmetric*, or *alternating*, *square*, respectively, of χ.

If $\{u_1, \ldots, u_n\}$ is a basis for U then $\{v_{ij} = u_i \otimes u_j : 1 \le i, j \le n\}$ is a basis for V. Then V_s and V_a have bases as follows:

$$V_s : \quad \{v_{ij} + v_{ji} : 1 \le i \le j \le n\},$$

$$V_a : \quad \{v_{ij} - v_{ji} : 1 \le i < j \le n\},$$

so their respective dimensions are $(n^2 + n)/2$ and $(n^2 - n)/2$. In order to compute χ_a say that $\widehat{S} = [s_{ij}]$. Then

$$T_a(x)(v_{ij} - v_{ji}) = S(x)u_i \otimes S(x)u_j - S(x)u_j \otimes S(x)u_i$$

$$\begin{aligned}
&= \sum_k s_{ki}(x)u_k \otimes \sum_l s_{lj}(x)u_l - \sum_k s_{kj}(x)u_k \otimes \sum_l s_{li}(x)u_l \\
&= \sum_{k,l}(s_{ki}(x)s_{lj}(x) - s_{kj}(x)s_{li}(x))(u_k \otimes u_l) \\
&= \sum_{k<l}(s_{ki}(x)s_{lj}(x) - s_{kj}(x)s_{li}(x))(v_{kl} - v_{lk}),
\end{aligned}$$

all $x \in G$. The "diagonal entries" are $s_{ii}(x)s_{jj}(x) - s_{ij}(x)s_{ji}(x)$, so

$$\chi_a(x) = \sum_{i<j}(s_{ii}(x)s_{jj}(x) - s_{ij}(x)s_{ji}(x)).$$

Doubling χ_a, we see that

$$\begin{aligned}
2\chi_a(x) &= \sum_{i \neq j}(s_{ii}(x)s_{jj}(x) - s_{ij}(x)s_{ji}(x)) \\
&= (\sum_i s_{ii}(x))(\sum_j s_{jj}(x)) - \sum_{i,j} s_{ij}(x)s_{ji}(x))
\end{aligned}$$

(the summands with $i = j$ cancel out). Thus

$$2\chi_a(x) = \chi^2(x) - \text{tr}(S(x)^2) = \chi^2(x) - \chi(x^2).$$

We have proved the next proposition.

Proposition 10.3.5 *If χ is an F-character of G, with symmetric and anti-symmetric squares χ_s and χ_a, then*

$$\chi_s(x) = \frac{1}{2}(\chi^2(x) + \chi(x^2)) \quad and \quad \chi_a(x) = \frac{1}{2}(\chi^2(x) - \chi(x^2)),$$

all $x \in G$. In particular, the function $x \mapsto \chi(x^2)$ is in $\text{Char}(G)$, since it is a difference $2\chi_s - \chi^2 = \chi^2 - 2\chi_a$ of characters.

If $x \in G$ set $\theta_2(x) = |\{y \in G : y^2 = x\}|$, the number of "square roots" of x in G. Clearly $\theta_2 \in \text{cf}(G)$, and $\text{Irr}(G)$ is a basis for $\text{cf}(G)$ by Corollary 5.2.19, so we may write

$$\theta_2 = \sum\{\nu(\chi)\chi : \chi \in \text{Irr}(G)\}$$

for some coefficients $\nu(\chi) \in \mathbb{C}$.

Proposition 10.3.6 *If $\chi \in \text{Irr}(G)$ then*

$$\nu(\chi) = |G|^{-1}\sum\{\chi(x^2) : x \in G\}.$$

Proof We have $\nu(\chi) = (\chi, \theta_2) = |G|^{-1} \sum_{y \in G} \chi(y)\theta_2(y)$. Fix y and observe that $\chi(y)\theta_2(y) = \sum\{\chi(x^2) : x \in G \text{ and } x^2 = y\}$. Then let y vary over G.   △

The function $\nu \colon \text{Irr}(G) \to \mathbb{C}$ is called the *Frobenius-Schur indicator*.

Theorem 10.3.7 (Frobenius and Schur) *If* $\chi \in \text{Irr}(G)$ *then* $\nu(\chi) = 1$, -1, *or* 0, *with* $\nu(\chi) = 0$ *if and only if* χ *is not real-valued.*

Proof By Proposition 10.3.5

$$\begin{aligned}
\nu(\chi) &= |G|^{-1} \sum_x \chi(x^2) \\
&= |G|^{-1} \sum_x \chi^2(x) - 2|G|^{-1} \sum_x \chi_a(x) \\
&= (\chi^2, 1_G) - 2(\chi_a, 1_G).
\end{aligned}$$

Suppose that $\chi \neq \overline{\chi}$. Then $0 = (\chi, \overline{\chi}) = (\chi^2, 1_G)$. But then also $(\chi_a, 1_G) = 0$, since χ_a is a summand of χ^2, and consequently $\nu(\chi) = 0$.

Suppose next that $\chi = \overline{\chi}$. Then $1 = (\chi, \overline{\chi}) = (\chi^2, 1_G)$. Thus $(\chi_a, 1_G)$ must be 0 or 1, and consequently $\nu(\chi) = 1 - 2(\chi_a, 1_G) = \pm 1$.   △

Corollary 10.3.8 *Suppose that* t *is the number of involutions in* G, *i.e.* $\theta_2(1) = t + 1$. *Then*

$$t + 1 = \sum\{\nu(\chi)\chi(1) : \chi \in \text{Irr}(G)\} \leq \sum\{\chi(1) : \chi \in \text{Irr}(G)\}.$$

Proof By definition $\theta_2 = \sum_\chi \nu(\chi)\chi$.   △

For example, recall that the dihedral group D_4 and the quaternion group Q_2 have identical character tables. If $G = D_4$ then the number t of involutions is 5, so $6 = \sum_i \nu(\chi_i)\chi_i(1) = 4 + 2\nu(\chi_5)$ and $\nu(\chi_5) = 1$. If $G = Q_2$ then $t = 1$, so $2 = \sum_i \nu(\chi_i)\chi_i(1) = 4 + 2\nu(\chi_5)$, and $\nu(\chi_5) = -1$. In particular the indicator is not in general determined just by the character table, although it is fairly easily calculated from the table provided that we also know the effect of the squaring map $x \mapsto x^2$ on the conjugacy classes.

Exercise

Show that

$$\sum\{\chi(1) : \chi \in \text{Irr}(S_n)\} \geq \sum_{k=0}^{\lfloor n/2 \rfloor} \binom{n}{2k} \frac{(2k)!}{2^k \cdot k!}.$$

Remark The inequality is actually equality, since $\nu(\chi) = 1$ for all $\chi \in \text{Irr}(S_n)$.

Proposition 10.3.9 *Suppose that G has even order and that t is the number of involutions in G. Then there is a real element $x \in G$, $x \neq 1$, with*

$$[G:C_G(x)] \le \left(\frac{|G|-1}{t}\right)^2.$$

Proof Label the characters in $\mathrm{Irr}(G)$ so that $\chi_0 = 1_G$; $\chi_1, \ldots, \chi_r$ are the nonprincipal real-valued characters; and $\chi_i \neq \overline{\chi_i}$ if $i > r$. Likewise label the conjugacy classes so that $K_0 = \{1\}$; $K_2, \ldots, K_r$ are the real classes; and the remaining classes are not real (see Proposition 10.3.2).

Then, as a slight refinement of Corollary 10.3.8, we have

$$0 < t \le \sum_{i=1}^{r} \chi_i(1).$$

Let $u = (1, 1, \ldots, 1)$ and $v = (\chi_1(1), \ldots, \chi_r(1))$ in $\mathbb{R}^r$. Applying the Cauchy-Schwarz inequality to the standard inner product on $\mathbb{R}^r$ we have

$$t^2 \le \left(\sum_{i=1}^{r} \chi_i(1)\right)^2 = (u, v)^2 \le \|u\|^2 \|v\|^2 = r \sum_i \chi_i(1)^2 \le r(|G|-1).$$

Multiply by $|G| - 1$ and divide by t^2 to obtain

$$|G| - 1 \le r\left(\frac{|G|-1}{t}\right)^2.$$

If $|K_i| > ((|G|-1)/t)^2$ for $1 \le i \le r$ then

$$\sum_{i=1}^{r} |K_i| > r\left(\frac{|G|-1}{t}\right)^2 \ge |G| - 1,$$

a contradiction, so there must be a real conjugacy class K_j with $|K_j| \le ((|G|-1)/t)^2$. Since $|K_j| = [G:C_G(x)]$ for any $x \in K_j$, the result follows. $\triangle$

Theorem 10.3.10 (Brauer and Fowler, 1955) *If $0 < n \in \mathbb{N}$ then there are only finitely many nonisomorphic simple groups G having an involution z with $|C_G(z)| = n$.*

Proof Given such a G, say that G has t involutions. Then $t \ge [G:C_G(z)] = |G|/n$, so $n > (|G|-1)/t$. By Proposition 10.3.9 there is an element $x \in G$ with $1 < [G:C_G(x)] < n^2$. The action of G on cosets of $C_G(x)$ thus gives an isomorphism (G is simple) from G into $\mathrm{Sym}(n^2 - 1)$, which of course has only finitely many subgroups. $\triangle$

This theorem of Brauer and Fowler, while not terribly deep, played an important role in the classification of finite simple groups (e.g. see [30]). When the Feit-Thompson Theorem [24] was proved it became clear that every (nonabelian) simple group *has* an involution, and the idea arose of classifying the groups in terms of centralizers of involutions – the basic idea being as follows. Take a simple group, or possibly some class of them, determine the centralizer of one of its involutions, usually taken in the center of a Sylow 2 - subgroup, and try to determine all of the finitely many simple groups having an involution with that same centralizer.

In many cases the simple group is determined up to isomorphism by the centralizer. That is the case, for example, for each $PSL(4, q)$ with $q \equiv 3$ (mod 4).

The classification also requires negative results – for example, Brauer and Suzuki proved in 1959 that no simple group has a (generalized) quaternion group as Sylow 2-subgroup.

In like manner various groups had to be ruled out as possible centralizers of involutions. However, in studying the case of a simple group G with $T \in$ $\mathrm{Syl}_2(G)$, z an involution in $Z(T)$, and $C_G(z) \cong \langle z \rangle \times \mathrm{Alt}(5)$, Z. Janko in 1965 was unable to prove nonexistence. Instead he proceeded to construct the first of the sporadic groups following the Mathieu groups. It is commonly called the *first Janko group* J_1 (he found more as well); it has order 175,560 and was constructed by Janko as a subgroup of $GL(7, 11)$ (see [42]). Some of the other sporadic groups arose in similar fashion.

See [39], pages 52–55, for some further interesting group theoretical applications of Corollary 10.3.8.

Exercise

If χ is a linear character of G, show that $\nu(\chi) = 1$ if and only if $\chi = \overline{\chi}$; $\nu(\chi) = 0$ if and only if $\chi \neq \overline{\chi}$.

We resume the discussion of χ^2 as the character of $T = S \otimes S$ on $V = U \otimes_F U$ that preceded Proposition 10.3.5.

The vector space V is isomorphic with the space of all $n \times n$ matrices over F via $v \mapsto A(v)$, where $v = \sum_{i,j} a_{ij} v_{ij}$ and $A(v) = [a_{ij}]$. If $x \in G$ then

$$
\begin{aligned}
T(x)v &= T(x) \sum_{i,j} a_{ij} v_{ij} = \sum_{i,j} a_{ij}(S(x)u_i \otimes S(x)u_j) \\
&= \sum_{k,l} (\sum_{i,j} s_{ki}(x) a_{ij} s_{lj}(x)) v_{kl},
\end{aligned}
$$

and so

$$
A(T(x)v) = \widehat{S}(x) A(v) \widehat{S}(x)^t. \tag{$*$}
$$

Exercise

Show that $A(v^*) = A(v)^t$, all $v \in V$.

Proposition 10.3.11 *Suppose that $\chi \in \mathrm{Irr}(G)$ is the character of representation S on U. If χ is real-valued then there is a matrix $A \neq 0$ such that $\widehat{S}(x)A\widehat{S}(x)^t = A$ for all $x \in G$. Any such matrix A is either symmetric or antisymmetric, and in fact $A^t = \nu(\chi)A$.*

Proof Since χ is real-valued, $(\chi^2, 1_G) = (\chi, \chi) = 1$. Thus 1_G has multiplicity 1 in χ^2, and it follows from Maschke's Theorem that $V = U \otimes U$ has a unique one-dimensional subspace V_0 for which $T(x)v = v$ for all $v \in V_0$, all $x \in G$. Thus we see by equation $(*)$ above that $\widehat{S}(x)A\widehat{S}(x)^t = A$ if and only if $A = A(v)$ for some $v \in V_0$.

Since $\chi^2 = \chi_s + \chi_a$ and $(\chi^2, 1_G) = 1$, we must have either $(\chi_s, 1_G) = 1$ (if $V_0 \subseteq V_s$) and $\nu(\chi) = 1$, or $(\chi_a, 1_G) = 1$ (if $V_0 \subseteq V_a$) and $\nu(\chi) = -1$ (see the proof of Theorem 10.3.7). Thus for each $v \in V_0$ we have $v^* = \nu(\chi)v$. Finally, apply the exercise preceding the proposition to see that if $v \in V_0$ then $A(v)^t = A(v^*) = A(\nu(\chi)v) = \nu(\chi)A(v)$. $\triangle$

Corollary 10.3.12 *If $\chi \in \mathrm{Irr}(G)$ is an $\mathbb{R}$-character then $\nu(\chi) = 1$.*

Proof We may assume that S is an $\mathbb{R}$-representation with χ as character. Set $B = \sum_{y \in G} \widehat{S}(y)\widehat{S}(y)^t$, and note that $B^t = B$ and $\widehat{S}(x)B\widehat{S}(x)^t = B$ for all $x \in G$. Furthermore each $\widehat{S}(y)\widehat{S}(y)^t$ has positive diagonal entries, so $B \neq 0$, and $B = B^t = \nu(\chi)B$; hence $\nu(\chi) = 1$. $\triangle$

The converse of Corollary 10.3.12 is true as well. The proof will require more work, however.

Recall that the *adjoint* of a complex (square) matrix A is the conjugate transpose of A; denote it by $A^\star$, so $A^\star = \overline{A}^t$. Note that, with the standard inner product for complex column vectors, $(Au, v) = (u, A^\star v)$ for all vectors u and v.

If $A^\star = A$ then A is called *Hermitian*, and if $A^\star = A^{-1}$ then A is called *unitary*. If $AA^\star = A^\star A$ then A is called *normal* – clearly Hermitian and unitary matrices are normal. It is a well-known fact from linear algebra (e.g. see [16] or [34]) that A is normal if and only if it is unitarily diagonalizable, i.e. there is a unitary matrix U so that $U^\star AU$ is diagonal. A Hermitian matrix A is called *positive definite* if $(Av, v) > 0$ for all nonzero vectors v; if A is positive definite then all its eigenvalues are positive real numbers.

A matrix representation U (over $\mathbb{C}$) of G is called a *unitary representation* if $U(x)$ is a unitary matrix for all $x \in G$.

Proposition 10.3.13 *Every $\mathbb{C}$-matrix representation T of G is equivalent with a unitary representation.*

Proof Set $A = \sum_{x \in G} T(x)^\star T(x)$ and note that A is Hermitian and positive definite. Since A can be unitarily diagonalized, it is easy to see that there is a positive definite Hermitian matrix B with $B^2 = A$. Observe that

$$(AT(x)A^{-1})^\star = A^{-1}T(x)^\star A = A^{-1}T(x)^\star \sum_y T(y)^\star T(y)$$

$$= A^{-1} \sum_y (T(x)^\star T(y)^\star T(y)T(x))T(x^{-1})$$

$$= A^{-1}AT(x^{-1}) = T(x^{-1}),$$

all $x \in G$. Set $U(x) = BT(x)B^{-1}$, all x. Then

$$U(x)^\star = (BT(x)B^{-1})^\star$$

$$= (B^{-1}AT(x)A^{-1}B)^\star$$

$$= B(AT(x)A^{-1})^\star B^{-1}$$

$$= BT(x^{-1})B^{-1} = U(x)^{-1},$$

and U is unitary.   △

Exercise

Use Proposition 10.3.13 to reprove Maschke's Theorem (5.1.1) in the case that $F = \mathbb{C}$. (If W is a T-invariant subspace show that its orthogonal complement $W^\perp$ is also T-invariant.)

Proposition 10.3.14 *If D is a diagonal matrix (over $\mathbb{C}$) then there is a diagonal matrix E with $E^2 = D$ and such that any complex matrix commutes with D if and only if it commutes with E.*

Proof Choose square roots $\beta_1, \ldots, \beta_m \in \mathbb{C}$ for the distinct diagonal entries $\alpha_1, \ldots, \alpha_m$ of D. By Lagrange Interpolation (e.g. [31], page 76) there is a polynomial $f(z) \in \mathbb{C}[z]$ with $f(\alpha_i) = \beta_i$, all i. Set $E = f(D)$.   △

Proposition 10.3.15 *Suppose that A is both unitary and symmetric (so $\overline{A} = A^{-1}$). Then $A = \overline{B}B^{-1}$ for some matrix B.*

Proof Choose U unitary so that $U^\star AU = D$, diagonal, and write $D = E^2$ as in Proposition 10.3.14. Then

$$D = D^t = (U^\star AU)^t = U^t A\overline{U} = U^t UDU^\star \overline{U} = \overline{U}^{-1}UDU^\star \overline{U},$$

so $U^\star \overline{U}D = DU^\star \overline{U}$, hence also $U^\star \overline{U}E = EU^\star \overline{U}$, and consequently $\overline{U}EU^\star = UEU^\star$. Since $\overline{D}D = D^\star D = I$ the diagonal entries of D are of absolute value 1, and likewise for those of E, so $\overline{E}E = I$ as well. Set $B = U\overline{E}U^\star$, and observe that

$$\overline{B}B^{-1} = \overline{U}E\overline{U^\star}U EU^\star = (UEU^\star)^2 = UDU^\star = A.$$

△

Theorem 10.3.16 *If $\chi \in \mathrm{Irr}(G)$ and $\nu(\chi) = 1$ then χ is an $\mathbb{R}$-character.*

Proof By Proposition 10.3.13, χ is the character of a unitary representation U. Since χ is real-valued $U \sim \overline{U}$ by Theorem 5.2.12, so there is
a matrix A with $A^{-1}U(x)A = \overline{U(x)}$, all $x \in G$. Thus $U(x)AU(x)^t =
A\overline{U(x)}U(x)^t = A(U(x)U(x)^\star)^t = A$, and consequently $A^t = \nu(\chi)A = A$
by Proposition 10.3.11. From $U(x)A = A\overline{U(x)}$ we see that $A^\star U(x)^\star =
\overline{U(x)}^\star A^\star$, or $A^\star U(x^{-1}) = \overline{U(x^{-1})}A^\star$, all x, and replacing x by x^{-1} we have
$A^\star U(x) = \overline{U(x)}A^\star$, all x. But then $AA^\star U(x) = A\overline{U(x)}A^\star = U(x)AA^\star$,
all x, so $AA^\star \in \mathcal{C}(U)$ and $AA^\star = \alpha I$ for some $\alpha \in \mathbb{C}$ by Proposition
5.1.4. Note that if v is any nonzero complex column vector of length 1 then
$\alpha = \alpha(v,v) = (AA^\star v, v) = (A^\star v, A^\star v) > 0$ in $\mathbb{R}$. Thus we may assume
that A has been replaced by βA, where $0 < \beta \in \mathbb{R}$ and $\beta^2 = 1/\alpha$, i.e. we
may assume that $AA^\star = I$, so A is unitary as well as symmetric. Then
by Proposition 10.3.15 we may write A as $\overline{B}B^{-1}$ for some B, and so from
$U(x)\overline{B}B^{-1} = \overline{B}B^{-1}U(x)$ we see that

$$B^{-1}\overline{U(x)}B = \overline{B}^{-1}U(x)\overline{B} = \overline{\overline{B}^{-1}U(x)\overline{B}},$$

all $x \in G$, and U is equivalent with a real representation.   △

Frobenius and Schur introduced the following terminology: χ in $\mathrm{Irr}(G)$ is
of the *first kind* if it is an $\mathbb{R}$-character, χ is of the *second kind* if it is real-valued
but is not an $\mathbb{R}$-character, and χ is of the *third kind* if it is not real-valued. If
χ is of the second kind and T is a $\mathbb{C}$-matrix representation with χ as character
then, as noted in the proof of Theorem 10.3.16, T must be equivalent with
$\overline{T}$, but T is not equivalent with a real representation.

The next theorem summarizes the results above.

Theorem 10.3.17 (Frobenius and Schur) *If $\chi \in \mathrm{Irr}(G)$ then*

1. *χ is of the first kind if and only if $\nu(\chi) = 1$,*
2. *χ is of the second kind if and only if $\nu(\chi) = -1$, and*
3. *χ is of the third kind if and only if $\nu(\chi) = 0$.*

Corollary 10.3.18 *If $\chi \in \mathrm{Irr}(G)$ then χ has Schur index 2 relative to $\mathbb{R}$ if
and only if χ is of the second kind, hence if and only if $\nu(\chi) = -1$.*

Exercise

Calculate the Frobenius-Schur indicator $\nu(\chi)$ for each $\chi \in \mathrm{Irr}(Q_m)$ (see
page 125). What are the Schur indices relative to $\mathbb{R}$?

Proposition 10.3.19 *If $\chi \in \mathrm{Irr}(G)$ and $\chi(1)$ is odd then χ is not of the
second kind, and hence χ has Schur index 1 relative to $\mathbb{R}$.*

Proof Suppose, to the contrary, that $\nu(\chi) = -1$. By Proposition 10.3.13 χ is the character of a unitary representation U, and by Proposition 10.3.11 there is a nonzero matrix A, with $A = -A^t$ and $U(x)AU(x)^t = A$ for all $x \in G$. As in the proof of Theorem 10.3.16, $AA^\star = \alpha I \neq 0$, so A is invertible. Thus $0 \neq \det A = \det A^t = \det(-A) = -\det A$, since the degree is odd, a contradiction. △

In a survey article [8] in 1963 R. Brauer cited a "rather old problem which is still open," namely that of determining by group theoretical properties of G the number of characters $\chi \in \mathrm{Irr}(G)$ of the first kind. Although answers have been obtained for a few special classes of groups, the problem seems to be still open in general.

Let us write $\nu^+(G)$, $\nu^-(G)$, and $\nu^0(G)$ to denote the number of $\chi \in \mathrm{Irr}(G)$ of the first, second, and third kinds, respectively, so

$$\nu^+(G) + \nu^-(G) + \nu^0(G) = c(G).$$

Thus Brauer's question asks for a determination of $\nu^+(G)$. Note that, by Proposition 10.3.2, $\nu^+(G) + \nu^-(G)$ is the number of real conjugacy classes of G, so of course $\nu^0(G)$ is the number of non-real classes. Burnside's Theorem (10.3.3) says that $\nu^+(G) = 1$ if $|G|$ is odd.

To conclude, we consider Brauer's question in the context of Frobenius groups. Recall that the characters of a Frobenius group have been described in Theorem 9.1.15.

Proposition 10.3.20 *Suppose that $G = M \rtimes H$ is a Frobenius group and that $1_M \neq \varphi \in \mathrm{Irr}(M)$.*

1. If $|H|$ is even then $\nu(\varphi^G) = 1$.

2. If $|H|$ is odd then $\nu_G(\varphi^G) = \nu_M(\varphi)$.

Proof (1) Choose an involution $z \in H$ and set $L = M \rtimes \langle z \rangle$, so $[L:M] = 2$. Then $\varphi^L|_M = \varphi + \varphi^z$ (e.g. by Theorem 6.4.1, or by an easy calculation) and $|M|$ is odd, so $\nu_M(\varphi) = \nu_M(\varphi^z) = 0$ by Theorem 10.3.3. Thus

$$\nu_L(\varphi^L) = |L|^{-1} \sum \{\varphi(x^2) + \varphi^z(x^2) + \varphi((xz)^2) + \varphi^z((xz)^2) \colon x \in M\}$$

$$= \frac{1}{2}[\nu_M(\varphi) + \nu_M(\varphi^z) + 1 + 1] = 1,$$

since $(xz)^2 = 1$ for all $x \in M$ by Proposition 9.1.11. Thus φ^L is an ℝ-character, and hence $\varphi^G = (\varphi^L)^G$ is an ℝ-character.

(2) Since $|H|$ is odd all elements of $G \setminus M$ have odd order. Hence if $x \notin M$ then $x^2 \notin M$. Also $\varphi^G|_M = \sum \{\varphi^h \colon h \in H\}$ (6.4.1 again) and $\varphi^G|_{G \setminus M} = 0$. Thus

$$\nu(\varphi^G) = |G|^{-1} \sum \{\varphi^h(x^2) \colon h \in H, \, x \in M\}$$

$$= |H|^{-1} \sum \{\nu_M(\varphi^h): h \in H\} = \nu_M(\varphi),$$

since $\nu_M(\varphi^h) = \nu_M(\varphi)$ for all $h \in H$. $\triangle$

Proposition 10.3.21 *Suppose that $G = M \rtimes H$ is Frobenius and that $\chi \in$ Irr(G), with $\chi|_H \in$ Irr(H). Then $\nu_G(\chi) = \nu_H(\chi|_H)$.*

Proof Calculate:

$$
\begin{aligned}
\nu(\chi) &= |G|^{-1} \sum \{\chi((xh)^2): x \in M,\ h \in H\} \\
&= |G|^{-1} \sum \{\chi(x \cdot {}^h x \cdot h^2): x \in M,\ h \in H\} \\
&= |M|^{-1}|H|^{-1} \sum \{\chi(h^2): x \in M,\ h \in H\} = \nu_H(\chi|_H)
\end{aligned}
$$

(recall that if $x \in M$ and $h \in H$ then $\chi(xh) = \chi(h)$ since $M \leq \ker(\chi)$). $\triangle$

We may use the propositions above to reduce Brauer's problem for a Frobenius group G to the corresponding problems for M and H.

Theorem 10.3.22 *Suppose that G is a Frobenius group with kernel M and complement H.*

1. If $|H|$ is even then

$$
\begin{aligned}
\nu^+(G) &= \frac{|M| - 1}{|H|} + \nu^+(H), \\
\nu^-(G) &= \nu^-(H), \\
\nu^0(G) &= \nu^0(H).
\end{aligned}
$$

2. If $|H|$ is odd then

$$
\begin{aligned}
\nu^+(G) &= 1 + \frac{\nu^+(M) - 1}{|H|}, \\
\nu^-(G) &= \frac{\nu^-(M)}{|H|}, \\
\nu^0(G) &= \frac{\nu^0(M)}{|H|} + c(H) - 1.
\end{aligned}
$$

Proof (1) Each character in Irr(H) extends to a character of the same kind in Irr(G) by Proposition 10.3.21. The kernel M is abelian, the $|M| - 1$ nonprincipal characters in Irr(M) lie in H-orbits of size $|H|$, and each of them induces to a character of the first kind on G by Proposition 10.3.20.

(2) All nonprincipal characters in $\mathrm{Irr}(H)$ are of the third kind by Theorem 10.3.3, and they remain so when extended to G. Of course 1_H extends to 1_G, of the first kind. The nonprincipal characters in $\mathrm{Irr}(M)$, lying in H-orbits of size $|H|$, induce to characters of the same kinds in $\mathrm{Irr}(G)$ by Proposition 10.3.20.   △

In many cases it is possible to answer Brauer's question for M and H fairly explicitly, since fairly complete information is available concerning the structure of Frobenius groups (see the previous chapter). For example, since M is nilpotent we have $M = K \times T$, with $|K|$ odd and T a 2-group, so $\nu^+(M) = \nu^+(T)$.

Exercise

Let $G = \mathrm{Aff}(F)$, where F is a finite field with $q > 2$ elements. Determine $\nu^+(G)$, $\nu^-(G)$, and $\nu^0(G)$ (see the exercise on page 176).

Bibliography

[1] M. Aschbacher, *Finite Group Theory,* Cambridge University Press, Cambridge, 1986.

[2] E. F. Assmus, Jr., and H. F. Mattson, Jr., *Perfect Codes and the Mathieu Groups,* Arch. Math. 17 (1966), 121–135.

[3] H. Bender, *A Group-Theoretic Proof of Burnside's $p^a q^b$-Theorem,* Math. Zeitschrift 126 (1972), 327–338.

[4] N. L. Biggs and A. T. White, *Permutation Groups and Combinatorial Structures,* Cambridge University Press, Cambridge, 1979.

[5] H. Boerner, *Representations of Groups with Special Consideration for the Needs of Modern Physics,* North-Holland, New York, 1970.

[6] R. Brauer, *On the Representation of a Group of Order g in the Field of the g-th Roots of Unity,* Am. J. Math. 67 (1945), 461–471.

[7] R. Brauer, *On Artin's L-series with General Group Characters,* Ann. Math. 48 (1947), 502–514.

[8] R. Brauer, *Representations of Finite Groups,* in *Lectures on Modern Mathematics I,* ed. T. L. Saaty, Wiley, New York, 1963.

[9] R. Brauer and K. Fowler, *Groups of Even Order,* Ann. Math. 62 (1955), 565–583

[10] W. Burnside, *Theory of Groups,* 2nd ed. (1911), Dover, New York, 1955.

[11] K. Chandrasekharan, *Introduction to Analytic Number Theory,* Springer-Verlag, New York, 1968.

[12] B. Char, K. Geddes, G. Gonnet, B. Leong, M. Monagan, and S. Watt, *Maple V Language Reference Manual,* Springer-Verlag, New York, 1991.

[13] A. H. Clifford, *Representations Induced in an Invariant Subgroup,* Ann. Math. (2) 38 (1937), 533–550.

[14] J. H. Conway, R. T. Curtis, S. P. Norton, R. A. Parker, and R. A. Wilson, *Atlas of Finite Groups*, Clarendon Press, Oxford, 1985.

[15] H. S. M. Coxeter and W. O. J. Moser, *Generators and Relations for Discrete Groups*, 4th ed., Springer-Verlag, Berlin, 1980.

[16] C. W. Curtis, *Linear Algebra*, Springer-Verlag, New York, 1984.

[17] C. W. Curtis and I. Reiner, *Methods of Representation Theory* (2 vols), Wiley Interscience, New York, 1981, 1987.

[18] C. W. Curtis and I. Reiner, *Representation Theory of Finite Groups and Associative Algebras*, Wiley Interscience, New York, 1962.

[19] N.J. deBruijn, *Pólya's Theory of Counting*, in *Applied Combinatorial Mathematics*, ed. E. F. Beckenbach, Wiley, New York, 1964.

[20] P. Diaconis, *Group Representations in Probability and Statistics*, Lecture Notes/Monograph Series, vol. 11, Institute of Mathematical Statistics, Hayward, CA, 1988.

[21] J. D. Dixon, *High Speed Computation of Group Characters*, Num. Math. 10 (1967), 446–450.

[22] L. Dornhoff, *Group Representation Theory*, Marcel Dekker, New York, 1971.

[23] W. Feit, *Characters of Finite Groups*, Benjamin, New York, 1967.

[24] W. Feit and J. Thompson, *Solvability of Groups of Odd Order*, Pacific J. of Math. (3) 13 (1963), 775–1029.

[25] G. Frobenius and I. Schur, *Über die Reellen Darstellungen der Endlichen Gruppen*, Sitzber. Preuss. Akad. Wiss. (1906), 186-208.

[26] D. M. Goldschmidt, *A Group Theoretic Proof of the $p^a q^b$ Theorem for Odd Primes*, Math. Zeitschrift 113 (1970), 373–375.

[27] D. M. Goldschmidt, *Group Characters, Symmetric Functions, and the Hecke Algebra*, American Mathematical Society, Providence, 1993.

[28] D. M. Goldschmidt and I. M. Isaacs, *Schur Indices in Finite Groups*, J. Algebra 33 (1975), 191–199.

[29] D. Gorenstein, *Finite Groups*, Harper and Row, New York, 1968.

[30] D. Gorenstein, R. Lyons, and R. Solomon, *The Classification of the Finite Simple Groups*, Mathematical Surveys and Monographs no. 40, American Mathematical Society, Providence, 1994.

[31] L. C. Grove, *Algebra,* Academic Press, New York, 1983.

[32] L. C. Grove and C. T. Benson, *Finite Reflection Groups,* 2nd ed., Springer-Verlag, New York, 1985.

[33] M. Hall, Jr., *The Theory of Groups,* Macmillan, New York, 1959.

[34] P.R. Halmos, *Finite-Dimensional Vector Spaces,* Springer-Verlag, New York, 1974.

[35] F. Harary and E. Palmer, *Graphical Enumeration,* Academic Press, New York, 1973.

[36] J. Humphreys, *Reflection Groups and Coxeter Groups,* Cambridge University Press, Cambridge, 1990.

[37] B. Huppert, *Endliche Gruppen I,* Springer-Verlag, Berlin, 1967.

[38] B. Huppert and N. Blackburn, *Finite Groups III,* Springer-Verlag, Berlin, 1982.

[39] I. M. Isaacs, *Character Theory of Finite Groups,* Academic Press, New York, 1976.

[40] G. James and A. Kerber, *The Representation Theory of the Symmetric Group,* Addison-Wesley, Reading, MA, 1981.

[41] G. James and M. Liebeck, *Representations and Characters of Groups,* Cambridge University Press, Cambridge, 1993.

[42] Z. Janko, *A New Finite Simple Group with Abelian 2–Sylow Subgroups,* Proc. Nat. Acad. Sci. 53 (1965), 657–658.

[43] D. L. Johnson, *Topics in the Theory of Group Presentations,* Cambridge University Press, Cambridge, 1980.

[44] D. L. Johnson, *Presentations of Groups,* Cambridge University Press, Cambridge, 1990.

[45] G. W. Mackey, *On Induced Representations of Groups,* Am. J. Math. 73 (1951), 576–592.

[46] E. Mathieu, *Mémoire sur l'Etude des Fonctions de Plusieurs Quantités,* J. Math. Pur. Appl. II sér. 6 (1861), 241–323.

[47] E. Mathieu, *Sur la Fonction Cinq Fois Transitive de 24 Quantités,* J. Math. Pur. Appl. II sér. 18 (1873), 25–46.

[48] J. Neubüser, *An Elementary Introduction to Coset Table Methods in Computational Group Theory,* in *Groups–St Andrews 1981,* ed. C. Campbell and E. Robertson, Cambridge University Press, Cambridge, 1982.

[49] P. M. Neumann, *A Lemma That Is Not Burnside's,* Math. Scientist 4 (1979), 133–141.

[50] D. S. Passman, *Permutation Groups,* Benjamin, New York, 1968.

[51] G. Pólya, *Kombinatorische Anzahlbestimmungen für Gruppen, Graphen und chemische Verbindungen,* Acta Math. 68 (1937), 145–254.

[52] J. H. Redfield, *The Theory of Group–Reduced Distributions,* Amer. J. Math. 49 (1927), 433–455.

[53] J. S. Rose, *A Course on Group Theory,* Cambridge University Press, Cambridge, 1978.

[54] J. J. Rotman, *The Theory of Groups,* 2nd ed., Allyn and Bacon, Boston, 1973.

[55] B. Sagan, *The Symmetric Group,* Wadsworth, Pacific Grove, CA, 1991.

[56] G. J. A.Schneider, *Dixon's Character Table Algorithm Revisited,* J. Symb. Comp. 9 (1990), 601–606.

[57] M. Schönert, *et al, GAP–Groups, Algorithms, and Programming,* ver. 3, Lehrstuhl D für Mathematik, Rheinisch–Westfälische Technische Hochschule, Aachen, 1995.

[58] J.-P. Serre, *Linear Representations of Finite Groups,* Springer-Verlag, New York, 1977.

[59] S. Sternberg, *Group Theory and Physics,* Cambridge University Press, Cambridge, 1994.

[60] M. Suzuki, *A New Type of Simple Groups of Finite Order,* Proc. Nat. Acad. Sci. USA 46 (1960), 868–870.

[61] M. Suzuki, *Group Theory I,* Springer-Verlag, Berlin, 1982.

[62] J. A. Todd and H. S. M. Coxeter, *A Practical Method for Enumerating Cosets of a Finite Abstract Group,* Proc. Edinburgh Math. Soc. (2) 5 (1938), 25–34.

[63] B. L. van der Waerden, *A History of Algebra,* Springer-Verlag, Berlin, 1985.

[64] K. S. Wang and L. C. Grove, *Realizability of Representations of Finite Groups,* J. Pure Appl. Algebra 54 (1988), 299–310.

[65] H. Weyl, *Symmetry,* Princeton University Press, Princeton, 1952.

[66] E. Witt, *Die 5-fach Transitiven Gruppen von Mathieu,* Abh. Math. Sem. Univ. Hamburg 12 (1938), 256–264.

[67] E. Witt, *Über Steinersche Systeme,* Abh. Math. Sem. Univ. Hamburg 12 (1938), 265–275.

[68] E. M. Wright, *Burnside's Lemma: A Historical Note,* J. Comb. Th., ser. B 30 (1981), 89–90.

[69] P. Yale, *Geometry and Symmetry,* Holden-Day, San Francisco, 1968.

[70] H. J. Zassenhaus, *The Theory of Groups,* Chelsea, New York, 1958.

Index

PURE AND APPLIED MATHEMATICS

A Wiley-Interscience Series of Texts, Monographs, and Tracts

Founded by RICHARD COURANT
Editor Emeritus: PETER HILTON
Editors: MYRON B. ALLEN III, DAVID A. COX, HARRY HOCHSTADT,
PETER LAX, JOHN TOLAND

*Now available in a lower priced paperback edition in the Wiley Classics Library.